A PLACE-BASED PERSPECTIVE OF FOOD IN SOCIETY

A Place-Based Perspective of Food in Society

Edited by

Kevin M. Fitzpatrick and Don Willis

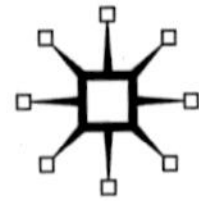

A PLACE-BASED PERSPECTIVE OF FOOD IN SOCIETY

First published in 2015 by
PALGRAVE MACMILLAN®
in the United States—a division of St. Martin's Press LLC,
175 Fifth Avenue, New York, NY 10010.

Where this book is distributed in the UK, Europe and the rest of the world, this is by Palgrave Macmillan, a division of Macmillan Publishers Limited, registered in England, company number 785998, of Houndmills, Basingstoke, Hampshire RG21 6XS.

Palgrave Macmillan is the global academic imprint of the above companies and has companies and representatives throughout the world.

Palgrave® and Macmillan® are registered trademarks in the United States, the United Kingdom, Europe and other countries.

ISBN: 978–1–137–40836–5

Library of Congress Cataloging-in-Publication Data

A place-based perspective of food in society / edited by Kevin M. Fitzpatrick and Don Willis.
pages cm
Includes bibliographical references and index.
ISBN 978–1–137–40836–5 (hardback)
1. Food—Social aspects. 2. Agriculture—Social aspects. 3. Space—Social aspects. I. Fitzpatrick, Kevin M.

GN407.P56 2015
394.1′2—dc23 2015006788

A catalogue record of the book is available from the British Library.

Design by Newgen Knowledge Works (P) Ltd., Chennai, India.

First edition: August 2015

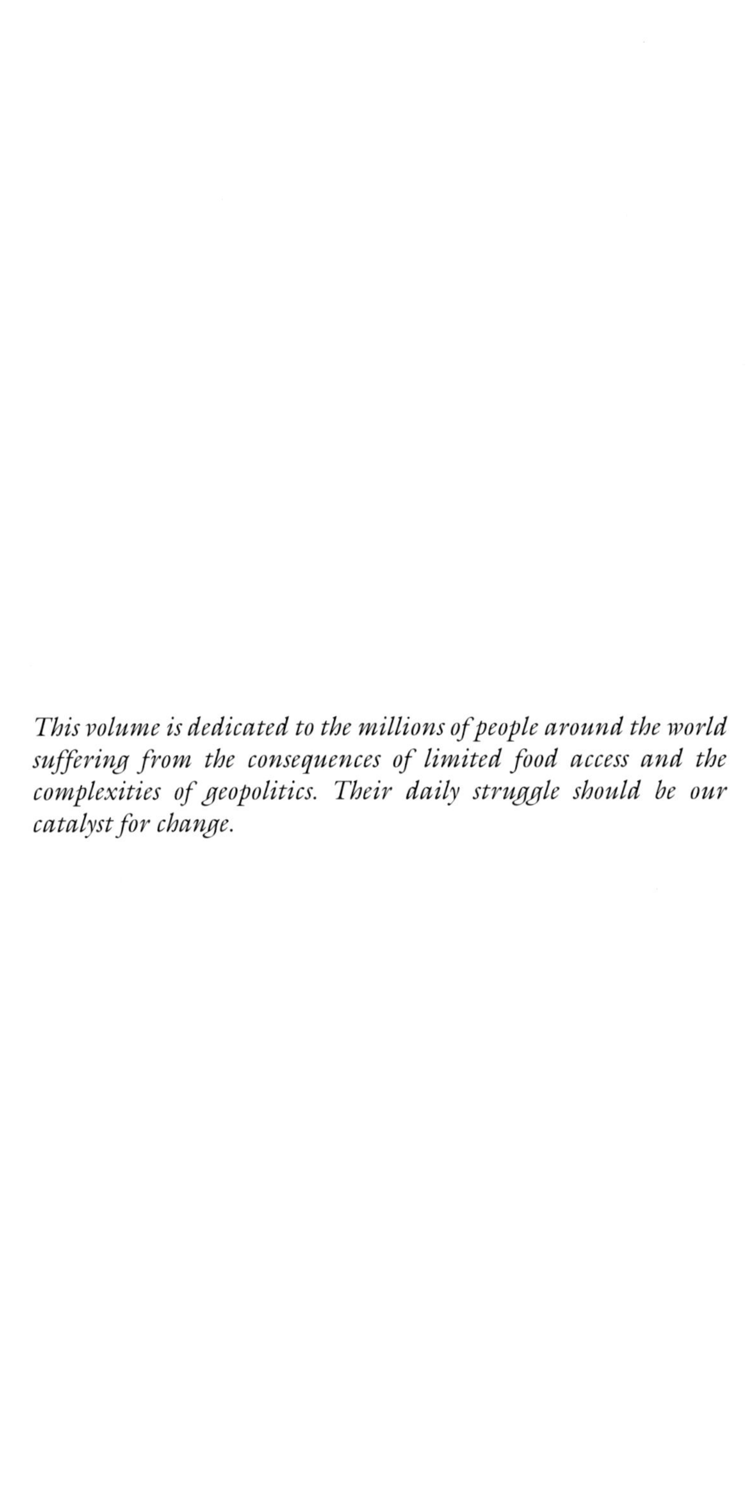

This volume is dedicated to the millions of people around the world suffering from the consequences of limited food access and the complexities of geopolitics. Their daily struggle should be our catalyst for change.

Contents

Illustrations

Figures

Tables

Acknowledgments

This volume represents an outgrowth of our continued interest in understanding the complexities of the food-place relationship. We acknowledge the support, assistance, insight, and thoughtful reflection of colleagues and staff in the development of this project. We would like to thank the University of Arkansas, Fulbright College of Arts and Sciences, the Department of Sociology and Criminal Justice as well as the University of Missouri, Department of Sociology, for creating the type of intellectual environments that allowed us to work on this project.

We would like to extend a special thanks to graduate assistants Stephanie Collier and Amanda Higginbotham whose work throughout this project was vital to its completion. In addition, Gail O'Connor and Dave Woodring did a fantastic job in carefully reviewing the manuscripts for content and style. Their constructive criticism improved the chapters and the overall organization of this volume. Finally, work like this cannot happen without the careful guidance and editorial expertise of our editorial assistant, Jocelyn Bailey. Jocelyn did a wonderful job of keeping everyone on task and carefully integrating these manuscripts into a coherent volume while providing expert advice to the authors and editors throughout the process. We would also like to thank the Palgrave Macmillan team—Nicola Jones, Elaine Fan, Deepa John, and Chelsea Morgan—for providing input and guidance throughout the editorial and production process.

Kevin M. Fitzpatrick: As with all my past work, I want to acknowledge my family and their support—particularly my wife Mary, who continues to be my biggest fan and critic. Don Willis: The support and guidance of others—namely, my wife Tara, close family and friends and, of course, my mentor, Kevin Fitzpatrick—have been invaluable to my work on this volume.

Introduction

Kevin M. Fitzpatrick and Don Willis

Place matters. It continues to be one of the primary social constants defining the distribution of standards of living and quality of life around the world. While place carries with it some conceptual baggage, there is mounting empirical evidence that zip code, county, country, and continent determine positional status in the world food crisis. Place plays a leading role in the story of what shapes our lives largely because of its intimate interplay with food and food-practices. Notwithstanding the broad distinctions between developed and undeveloped/underdeveloped food systems around the world, place continues to create barriers in the production and consumption of food for the world's populations. Even at the local level, as highlighted in the *New York Times* article "Our Coming Food Crisis" (G. Nabhan, July 21, 2013), place matters when climate changes, and farmers are forced to adapt their production strategies to accommodate dramatic shifts in temperature and rainfall.

We, of course, are not confronting just an American food crisis; farmers around the world are facing a host of problems compounded by scarce resources, an imbalance in the supply-and-demand equation, current food policies, climate change, and social, economic, cultural, and political factors that are unique to particular places. Whether focusing efforts to eliminate food deserts throughout the United States, or addressing the complexities farmers face in trying to deal with the problems of food production, distribution, and scarcity in sub-Saharan Africa, we see how the impact of place can be both dramatic and multifaceted.

Besides being places, spaces are geographic units with physical, cultural, social, and economic properties, as well as personally defined places. We would argue that both aspects of place matter for understanding the way that we interact with food and place. For some segments of the world population, being in the wrong place at the wrong time is not just a matter of bad luck but also the result of social structure. Resources are structurally distributed across social landscapes; it is not by accident that low-income racial and ethnic minorities living in urban areas suffer high rates of obesity, malnutrition, heart disease, and food insecurity. These often highly segregated spaces achieve and perpetuate an "ecology of disadvantage" that is isolated from more privileged places and populations. There are countless examples around the world illustrating this very notion of place-based limits that compound problems many populations face when trying to access food.

As our discussion suggests, context can be as simple as geographic location, but it can be as complicated as multidimensional layers of risk (hazards) and resources (social capital and networks) that are distributed unequally across population subgroups. Simply, or from a more complicated viewpoint, food and its role both in ancient and in modern societies, was and still is highly dependent on place. Our motivation in designing this particular volume has been the lack of any comprehensive treatise of the food-place nexus. In addition, we are excited about providing a platform for scientists across multiple disciplines (sociology, anthropology, history, economics, political science, health sciences, etc.) to address this often-overlooked component in exploring the twenty-first-century story of the world food crisis, how we got here, and where we might go next.

In an article by Feagan (2007), he provides a useful overview that highlights the importance of place in understanding food systems; community-supported agriculture, community food security, and sustainable communities are all important examples of ongoing strategies that advance our interpretation of the food-place relationship. Understanding the context of food and its impact on the world's population is a "glocal" problem in need of innovation, political vision, and a new approach to solution-building that demands cross-disciplinary conversations.

This volume is organized into four distinct parts. The first part, "Historical Contexts," begins with a comprehensive discussion of the history of food and agriculture. Roudart and Mazoyer take on the formidable task of providing insights into the origins and propagation of agriculture from the Neolithic Era to the present day. Their discussion formulates an interesting interrelationship with this history and the evolution of our contemporary diet, which has constantly been modified in the context of place. The second chapter in this part addresses the industrialization and globalization of food. Maya-Ambía frames this discussion in the context of two important frameworks (Food Regime and Global Chains), with insightful observations from his own work and travel. Despite the big business of food, Maya-Ambía identifies key elements regarding the importance of the "local" place and its complicated culture and structure impacting food distribution and consumption. The third chapter in this part offers a set of historical examples that highlight the importance of the food-place relationship. Tauger walks us through history and around the globe while teaching us important lessons from the past—arguing that ignoring the role of place can have disastrous consequences for populations if care is not taken regarding overall food security while minimizing risk.

The second part of the volume, "Social and Cultural Contexts," begins with an exploration of the role of knowledge and food movements in our understanding of the food-place relationship. Sumner highlights how we acquire knowledge and use this place-based information to think more deeply and act more responsibly about what we grow, how we grow it, and, ultimately, how we consume it. The remaining two chapters in this second part of the volume provide fascinating insights into the distinctiveness of place-based food and flavor, using examples of Southern food and the Brooklyn "culinary renaissance." Byrd explores the intersection of food, space, and structural inequality in the South from a historical as well as a contemporary viewpoint. Her comments regarding this complicated interrelationship are fresh and insightful as Southern food and cookbooks anchor this exploration into foodways. The final chapter in this part, authored by LeBesco and Naccarato, explores the emerging food movement in Brooklyn and its role in ethnic and immigrant communities. The paradox of old and new is a food-related tension, not unlike the one found in other places around the country, yet LeBesco and Naccarato's observations on these neighborhoods are fascinating and peppered with interviews and insights from the cast members in this complicated drama.

The third part of the volume, "The Context of Power and Inequality," begins with Wengle's discussion of the political economy of food and a cross-national comparison of food policy in the United States and in Russia. We learn an important lesson in this chapter about the distinct role of place and political context in the formation and function of food-related policies. The analysis illustrates the uneven effects of national policy within and across nations. The chapter also raises a number of critical questions for future research regarding how significantly food-related policies have perpetuated a disappearance of the "agriculture of the middle," which has implications for the connection of farming practices to place and social context. In the following chapter, Larimore and

Schmutz explore the power dynamics of place within a global food system and the manner in which they reproduce inequalities both in food access and in environmental degradation. The authors begin to connect the dots between food-related inequalities and intersecting social inequalities of race, class, and gender. In doing so, a more complex picture of place-based solutions comes into view—one that highlights social inclusion over solely economic concerns. Gartin completes this part of the volume with a chapter that takes us across three continents to examine the most intimate of concerns regarding the food-place nexus—health and well-being. Gartin uses each of these three places as case studies and analyzes the tensions between them and the global food market. These case studies provide examples of varying solutions to this tension and their outcomes in terms of access, diet, and health.

The fourth and final part of the volume, "The Future of Food," is a discussion of solutions: what is working and what is not. This part begins with Timmer's chapter on place-based responses to the global food economy. The chapter explores the connection between global and local systems of food distribution and access while focusing on the important role that the "food revolution" is playing in reclaiming the importance of place in our understanding of where food comes from and its impact on population health. Lafferty follows with a chapter that draws on scholarship that addresses "troubling" alternative food practices and pushes the conversation further by asking whether food movements are reaching the places and spaces where access to healthy, culturally appropriate food is needed most. While some earlier chapters have critiqued the market and economic focus of many alternative food movements, Lafferty reframes the discussion in terms of social justice and food sovereignty. In the final chapter of this volume, Hossfeld and colleagues highlight place by focusing their lens on Southeastern North Carolina and a particular local food initiative there. In describing a single initiative and its operation within a unique place, the authors provide a rich and detailed description of the successes and struggles faced by such a program—often from the perspective of farmers who have participated in the initiative. The authors argue that this program's success in access, equity, and inclusion make it an exemplary model for other local initiatives, though they are also quick to point out how there is no single solution to such a complex web of inequalities.

Food and place are ubiquitous in the lives of all humans. They are fundamental to our experience of the world both as a physical and a social environment; thus, these concepts provide fertile ground for work from a multitude of scholarly disciplines and approaches. The story of humans and their role within the relationship between food and place is a complicated one. This volume represents an interdisciplinary effort to tell that story. While these chapters do a masterful job of outlining the importance of the food-place relationship across space, time, country, region, and discipline, the story is by no means complete. Advocates and researchers must remain vigilant in their efforts to bring both theoretical understanding and real-world praxis to the issues these authors have discussed. Our hope is that this volume has begun a process of cross-fertilization of knowledge and ideas that will lead to the type of innovative thinking and action needed to address the complexities and consequences of the food-place relationship.

References

Feagan, R. (2007). The place of food: Mapping out the 'local' in local food systems. *Progress in Human Geography*, 31(1), 23–42.

Nabhan, G. P. (2013, July 21). Our coming food crisis. *The New York Times*. Retrieved from http://www.nytimes.com/2013/07/22/opinion/our-coming-food-crisis.html?_r=0

Part I

Historical Contexts

Chapter 1

The Origins and Propagation of Agriculture: The Formation of the Contemporary Diet

Laurence Roudart and Marcel Mazoyer

Introduction

The human species, *Homo sapiens*, is heterotrophic, meaning that to live, human beings must consume organic matter (carbohydrates, lipids, proteins, and nucleic acids) provided by other living entities, whether plants, animals, fungi, or microorganisms. To subsist in a given place, a human population must accordingly be in a position to continually procure edible organic matter. These foodstuffs may be produced on-site or imported. Thus, whatever the era, the presence of a human population in a given place is determined by the edible species available on-site and by the knowledge of techniques (implements and practices) for deriving foodstuffs from these species. But it is also determined by foods produced in other places and by the means of transporting them to their place of consumption. In short, the presence of a human population depends both on ecological and on technical and cultural conditions.

From an ecological standpoint, each place of production can be characterized by its ecosystem—that is, its biotope (climatic and soil conditions) and its biocenosis (all the plant and animal populations living in the place). An ecosystem can be modified by human activity. During every period, the ecosystem, whether modified or not, determines the range and the proportions of the different species, especially food-providing species, that are present. Culturally speaking, for the last 200,000 years, the human species has acquired its nourishment from hunting, fishing, or the gathering of wild species living spontaneously in ecosystems that have barely, if at all, been modified by humans. Diets were entirely dependent upon the food-providing species present in each place, with most groups of humans being nomadic. Only in the Neolithic period, between 10,000 and 5,000 BP (Before Present, conventionally taken to mean before 1950) did a few recently settled societies of hunters, gatherers, and fishers develop of their own accord into communities of farmers, who grew plants and bred animals to produce food.

Today, the human race obtains virtually all its food from forms of agriculture, which vary greatly from place to place. For instance, a farmer in the Corn Belt of North America, operating on his own with powerful machines, can cultivate several hundred hectares of corn and produce thousands of tons of grain for conversion into ethanol. He functions in a very different ecological, technical, economic, and social context from that of his counterpart in the inner Niger delta in West Africa, who—with the help of his family and draft zebus—can cultivate a few hectares of rice and produce fewer than ten tons of grain that are intended mainly for his family's consumption.

Basic foodstuffs, being heavy, bulky, and perishable, are ill-suited to transport over long distances. Yet in ancient times, cities of several thousand inhabitants in the Nile Valley and in Mesopotamia were supplied by water transport. Rome, which reached about a million inhabitants in 2000 CE, was provisioned by boats bringing grain from all parts of the Mediterranean basin. Until the advent of the railways in the nineteenth century and of motorized vehicles in the twentieth, land transportation of foodstuffs was effected by cart, pack animals, or human backs; and it took place only over short distances and in limited volumes. Even today, when the means of transport have never been so powerful, the international trade in agricultural produce accounts for only about 15 percent of world output (FAO, 2014). Farmers and their families, accounting for some 40 percent of the world's population, continue to consume a major part of their own crops. And in developing countries, the majority of the nonagricultural rural population and many town-dwellers continue to consume what is produced in their local region or country.

Although the emergence of agriculture is a very recent phenomenon when compared to the vast duration of human prehistory, it is nonetheless too ancient for any written testimony or memory to have survived. Over the last few decades, however, advances in archaeology have enabled science to deduce with reasonable accuracy which places, at what time, and under which circumstances societies of predators became societies of farmers. Archaeology also has created knowledge of where, when, and how Neolithic farmers subsequently spread cropping and breeding practices across the world. Lastly, research conducted in other fields, such as history, anthropology, geography, or agronomy, have established that it is now possible to reconstitute the way in which the main types of Neolithic farming and diets continued to develop and diversify in relation to the ecological and cultural conditions pertaining to each place in the world. These are the central issues we address in this chapter.

Ecosystem Diversity at the End of the Paleolithic Era

At the end of the Paleolithic era, some twelve thousand years ago, following the long Würm glaciation, the Holocene, a relatively warm period, commenced and has, with some slight variations, lasted until the present. The ice cap retreated by three thousand kilometers, and new vegetal landscapes came into being (see Figure 1.1) (Cox & Moore, 2010). The humans of the time adapted to these diverse ecological conditions. They improved their chipped stone tools and shaped highly specialized new tools from materials such as wood, ivory, or horn, allowing them to exploit the particular resources of each environment. With the combined effects of global warming, the global dissemination of vegetal or animal populations with edible species, and the development of appropriate modes of predation, the human population was close to reaching the area it occupies today. This area extends from the southernmost point of the South American continent, home to the now-extinct Phrygian people, to the Arctic polar regions inhabited by the Eskimos; and it goes from sea level up to the grasslands of Central Asia and the Andes, at altitudes of five thousand meters.

During this time, most hunters, fishers, and gatherers moved from one encampment to the next after exhausting local resources. However, in some especially favored places that were rich in food species, certain groups of people could settle each year during the harvest season—or even year-round—at seaside or lakeside locations that were rich in fish or other seafood. Certain groups could even become sedentary, thanks to progress made in conservation processes (drying, smoking, cold storage, silos, etc.) (Guilaine, 1991; Testart, 2012). Under these conditions, a few millennia after the beginning of the Holocene, several human societies worked out a new form of toolmaking—stone polishing. The Neolithic era then began, and with it the earliest developments of

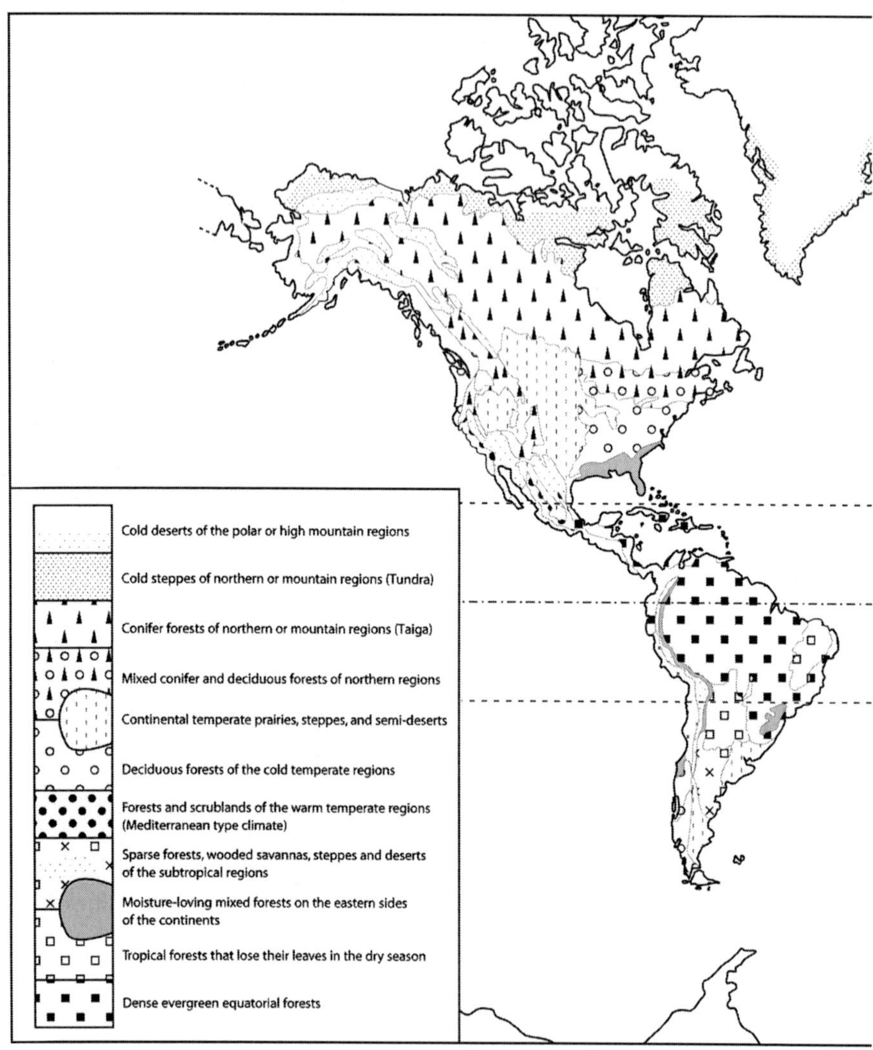

Figure 1.1 Schematic Map of "Original" Plant Formations 10,000 Years Ago

Source: Mazoyer M., & Roudart L. (2006). *A History of World Agriculture: From the Neolithic Age to the Current Crisis.* New York, NY: Monthly Review Press.

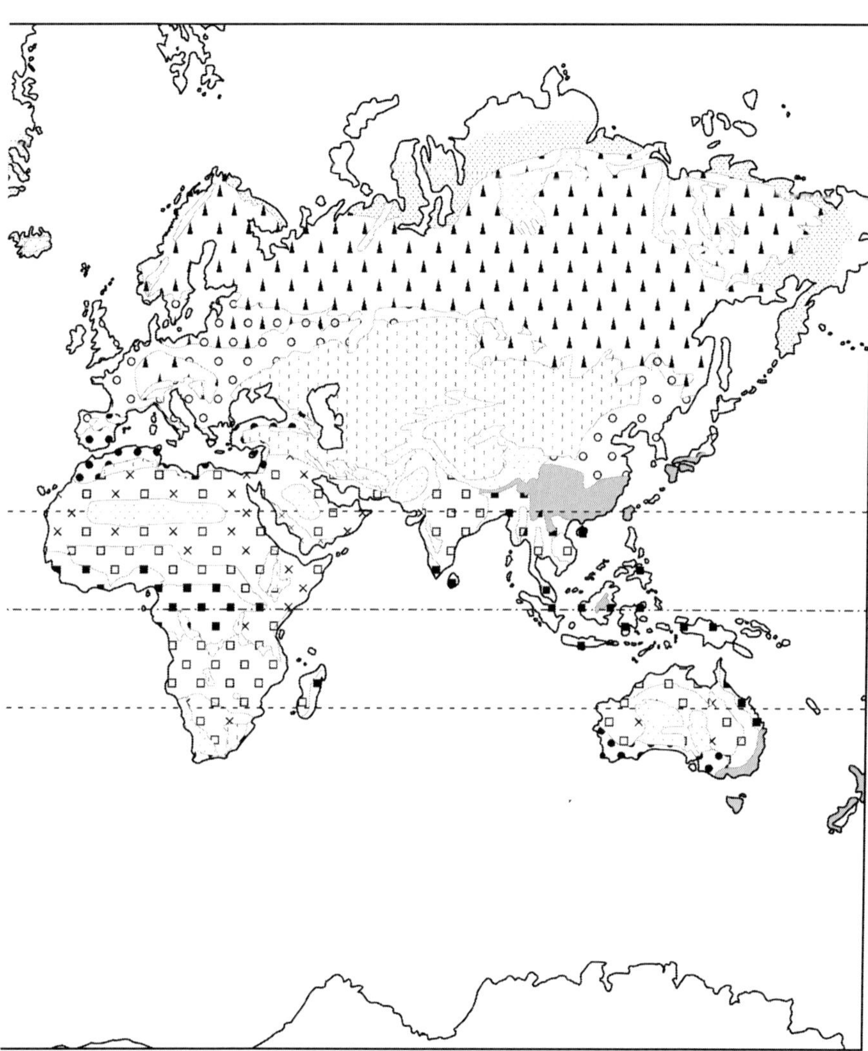

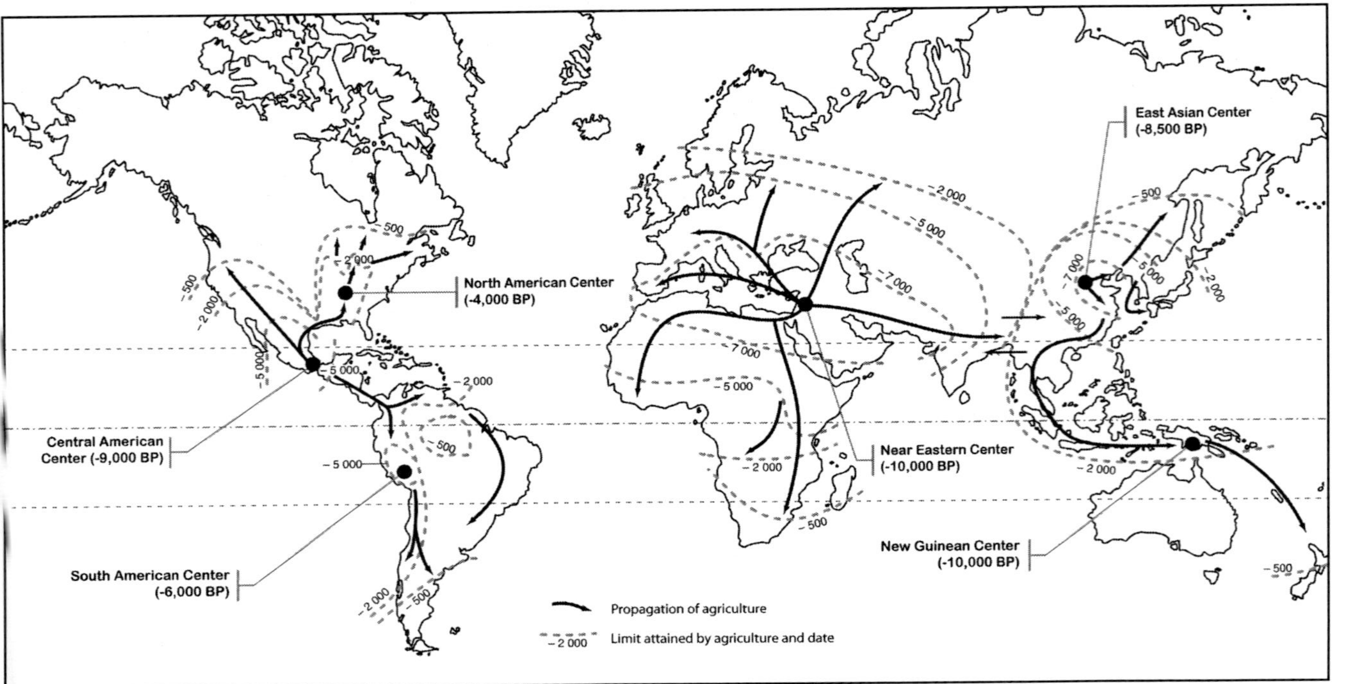

Figure 1.2 Centers of Origin and Areas of Propagation of the Neolithic Agricultural Revolution

Source: Mazoyer M., & Roudart L. (2006). *A History of World Agriculture: From the Neolithic Age to the Current Crisis*. New York, NY: Monthly Review Press.

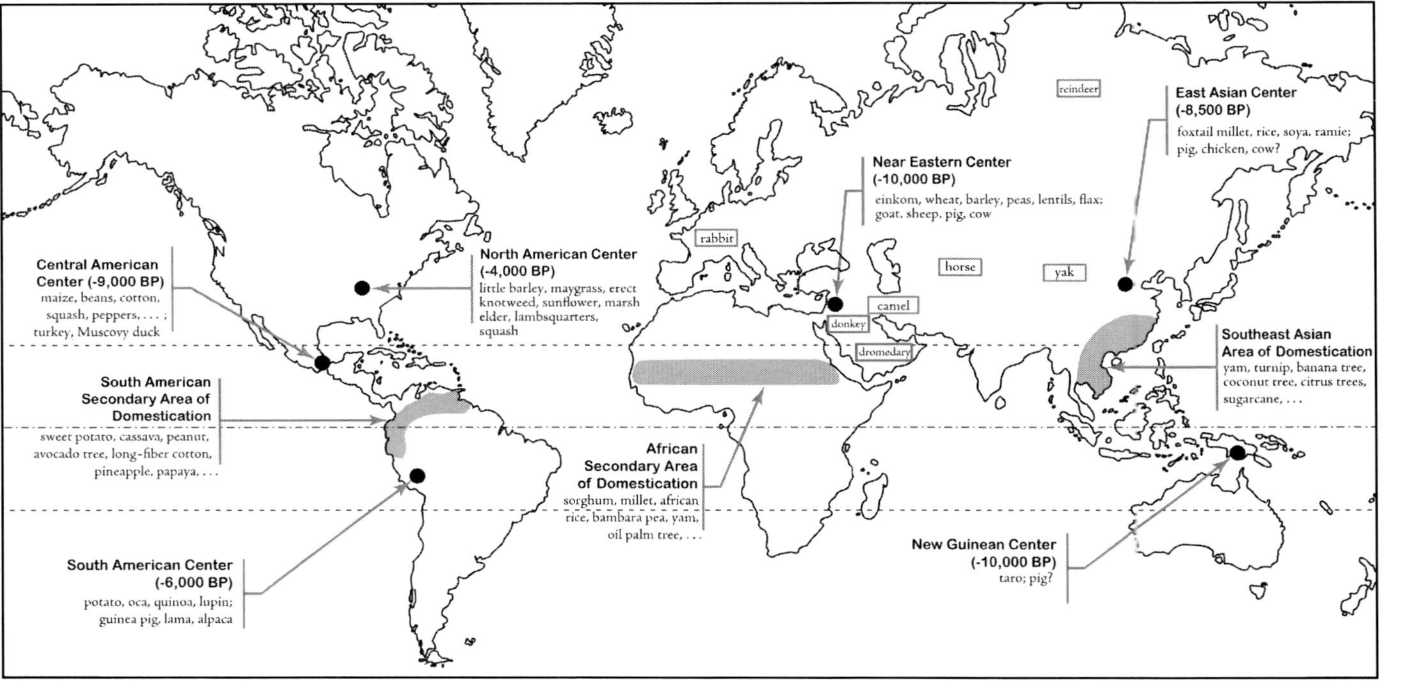

Figure 1.3 Centers of Origin of the Neolithic Agricultural Revolution and Secondary Areas of Domestication

Source: Mazoyer M., & Roudart L. (2006). *A History of World Agriculture: From the Neolithic Age to the Current Crisis*. New York, NY: Monthly Review Press.

agriculture. The emergence of Neolithic agriculture was long regarded merely as a sort of rapidly spreading innovation that was made necessary by an acute food crisis resulting from the overexploitation of wild resources. However, in recent decades, archaeological research has shown this to be untrue. The transformation from a society subsisting on simple predation into a society living mainly on the product of cultivation and breeding is to be seen as a complex chain of material, social, and cultural changes that condition one another and unfold over several hundreds of years (Demoule, 2009a).

Emerging Agriculture and the Primary Food Patterns in the Neolithic Era

Archaeological research has shown that agriculture appeared independently in various parts of the world, which we call the "places of emergence" or "centers of origin" of Neolithic agriculture. Given the present state of knowledge, the existence of six such places of emergence can be identified (see Figures 1.2 and 1.3): *the Near Eastern* place, which formed in Syria-Palestine between 10,000 and 9,000 BP; the *Central American* place, which developed in Southern Mexico after 9,000 BP; the *East Asian* place, which formed in Northeast China on the middle reaches of the Yellow River after 8,500 BP, before extending to the northeast and the southeast; the *New Guinean* place, which perhaps emerged in the interior of Papua-New Guinea some 10,000 years ago; the *South American* place, which developed in the Andes before 6,000 BP; and the *North American* place, which grew in the middle basin of the Mississippi between 4,500 and 2,500 BP. A number of other regions in Africa and Asia were perhaps also places of emergence, but the issue continues to be debated (Demoule, 2009a; Grigg, 1977; Mazoyer & Roudart, 2006).

In the *Near East*, and more exactly in the triangle formed by the Dead Sea, the Taurus range, and the Zagros Mountains, the best known of the six places of emergence of Neolithic agriculture can be found. It is here that we can document the slow transition from predation to agriculture that lasted over a thousand years (Cauvin, 2007).

Near Eastern Place of Emergence

Abundance of Resources, Sedentary Populations, and Tool Specialization

In the Near Eastern region, the global warming of the Holocene period gradually replaced the cold steppe, with artemisia and reindeer—with a savanna that was rich in wild food resources such as acorns and pistachios, cereal grains (einkorn, wheat, barley) and legumes (peas, lentils, grass peas, vetch), game (boars, aurochs, onagers, mountain sheep, wild goats, deer, gazelles, rabbits, hares, birds), and fish in some places. Resources were even abundant enough for the human population to settle. Living first in grottos, this population increased gradually. Once the grottos became insufficient, people moved into artificial habitats, composed of round dwellings within wooden superstructures, grouped in villages of 0.2 to 0.3 hectare (Cauvin, 2007). Several tools and techniques determined the use of the new resources and the development of the new sedentary lifestyle: axes and adzes of polished stone for cutting and shaping the timber, stone sickles for harvesting, stones and rollers for grinding grains, and silos for stocking wild cereals. Afterward, for thousands of years, these tools and techniques were used in farming.

Protoagriculture, Protobreeding, and Domestication

Protoagriculture and protobreeding are the terms we give to the earliest cultural and breeding practices applied to still-wild populations of plants and animals. The very few remaining archaeological traces of these practices allow us to surmise that the earliest sowing practices occurred near the habitations, on terrains already cleared and enriched with domestic waste. Protoagriculture might also have developed on land enriched by alluvial deposits from overflowing rivers, where appropriate, and later on wooded ground: indeed, with axes made of polished stone, it was fairly easy to fell trees, which were then burned before cropping began (Rollefson & Köhler-Rollefson, 1992).

In other words, in the Neolithic era, sedentary human societies subjected small wild populations of plants and animals to new, *artificial* conditions, resulting from the practices of protoagriculture. These populations thereafter led a separate existence, distinct from that of their wild fellows. After several generations, many of these plant and animal populations proved amenable to domestication—losing some of their genetic, morphological, or behavioral features that had become incompatible with their new existence—and acquired other, more advantageous features. Although they continued to resemble the wild populations from which they had sprung, these domestic populations were now distinguished by a series of characteristics that formed the "domestication syndrome." For instance, wild cereals have ears that are easily detached and shelled, thus favoring their natural dissemination, while the ears of domesticated cereals are difficult to detach and shell, which is conducive to their multiplication by harvesting and sowing (Pernès, 1983). In the Near East, the earliest traces of completely domesticated plants and animals (see Figure 1.3) date from 9,500 BP. For these plants and animals to have been domesticated by that time, the protoagriculture and protobreeding of their still-wild forms must have begun several dozens, if not hundreds, of years before (Gautier, 1990; Harlan, 1992).

The period between 9,500 and 9,000 BP also saw the move away from small villages, with their well-spaced round dwellings, to larger villages with more contiguous rectangular houses, and populations ten times as large. There were also major developments of polished stone axes and adzes, utilitarian fired pottery, and a large number of feminine statuettes and figurines, probably symbolizing fertility. It is not easy to establish links of cause and effect between these innovations since they do not appear in chronological order in the different excavated sites. However, they are all situated in the entire Near Eastern area by 9,000 BP, at a time when domesticated plants and animals were providing the human population with the bulk of their food (Cauvin, 2007).

These changes attest to a major increase in the population as well as to far-reaching social and cultural transformations that are difficult to chart accurately. However, the evidence suggests that small domestic groups that were involved in production and consumption came into being, each with its own roof, fire, and silo. For each of these groups, it would not be difficult to sow its preferred seeds on a prepared patch of ground; nor would it be difficult to capture, tame, and reproduce in captivity the most easily mastered of its favorite types of game. Even hunter-gatherers were capable of doing this. What was no doubt more difficult was to preserve the harvest, produced by their own seeds from the pre-existing "gathering rights" of the other groups, and to subtract the animals one had raised from their "hunting rights."

It was also difficult to keep a part of the harvest as seeds and a part of the herd for reproduction purposes and to then distribute the fruits of their labor between the different members of the group—not only on a daily basis, but also when the eldest members died or the group was subdivided. Thus, these domestic groups, which had now become farmers, certainly respected a minimum number of new social rules allowing their own reproduction as well as the proportional reproduction of the cultivated plants and domestic animals upon which their survival depended (Demoule, 2009a; Liu, 2007; Mazoyer & Roudart, 2006). In addition, one may assume that this new lifestyle could not have been understood, transmitted, and improved upon without the help of a renewed form of language that could describe the new physical conditions (habitat, environment, tools, farming practices), the new social rules, and the corresponding ideas, representations, and beliefs (Bellwood, 2004; Cavalli-Sforza, 2001; Renfrew, 1990).

Conditions for the Emergence of Agriculture

The places where agriculture emerged in Neolithic times have in common a range of ecological and technocultural characteristics, which were seldom conflated. First, the wild, food-producing species, notably plants, were plentiful enough to enable small groups of hunters, gatherers, and fishers to become sedentary and to subsist for generations. That was long enough for the populations to perfect techniques for building solid, perennial dwellings and to develop sophisticated techniques for harvesting, preserving, and preparing foodstuffs. Second, some of the local, wild, food-producing species could be successfully grown, bred, and domesticated to satisfy a major part of the population's dietary requirements. Third, the societies then living in such places were in a position to establish the rules of conduct by which agricultural activities could be pursued. In short, agriculture could emerge only in places that were ecologically privileged and, at the same time, particularly advanced in a cultural sense.

In these especially favorable places, the emergence of agriculture was not a response to some acute crisis in predation resulting from the overexploitation of wild resources by a rapidly growing sedentary human population. There is in fact no evidence of any such crisis during the lengthy transition period between predation and agriculture (Cauvin, 2007). Nor was it caused by the spread of some belated and fortunate discovery that one could reap more abundantly what one had sown. After all, it is scarcely imaginable that *Homo sapiens* gatherers, who had been sedentary for centuries and had observed how seeds sown serendipitously around their dwellings produced easily harvested ears of grain or leguminous pods, had remained unaware of the connection between sowing and harvesting. As for capturing, taming, and raising young animals, any hunter-gatherer's child was capable of doing so. The dog, prized as the nomadic hunter's asssistant, had been domesticated since the late-Paleolithic period.

Most plausibly, the populations in these places developed cropping and breeding practices once the demographic density had attained a critical threshold. Beyond this threshold, wild food resources being plentiful but limited, the working time devoted to gathering and hunting increased to the point of becoming dangerously close to the total working time available. It then became worthwhile to develop cropping and breeding, which allowed the time spent in gathering plants and capturing animals to be reduced. This entailed an increase in the working time needed and the total time spent on work annually (Sahlins, 1974). This appears to be the ultimate reason why societies of settled hunter-gatherers developed crop-growing and animal breeding. The emergence of agriculture was therefore the consequence of a series of demographic and technocultural developments that were achieved by populations living in new, exceptional, and particularly propitious ecological circumstances (Harlan, 1992; Mazoyer & Roudart, 2006).

Population Growth and New Diets

Many archaeologists believed until recently that the Neolithic agricultural revolution had led to improved diet and health, followed by increases in life expectancy and population (Skorupa, 2013). The fact that population increased is not disputed. But recent work on the Near Eastern place of emergence shows that things are not as simple as they appear. On the one hand, living in crowded conditions in the new villages facilitated the transmission of contagious diseases; so mortality increased. On the other hand, the sedentary lifestyle curtailed the interval between pregnancies; so the birth rate rose much faster than the death rate, resulting in a rapid growth in population (Bocquet-Appel & Bar-Yosef, 2008). Nevertheless, recent analyses of human remains have shown that in some places, the Neolithic agricultural revolution was followed by a worsening health situation (Skorupa, 2013). Indeed, the human diet lost diversity because it comprised more and more food from cropped or bred species and less and less food from wildlife. This is how several major primary dietary types were formed in the Neolithic, on the basis of the following plants: wheat, barley, peas, and lentil in the Near East; maize and beans in Central America; potato, oca, and quinoa in the Andes (Harlan, 1992); sunflower, little barley, millet, and erect knotweed in North America (Smith, 2007); rice, millet, and soya in East Asia (Jiang & Liu, 2006); and taro in Papua (Bulmer, 1975). Food could then be cooked as in the Paleolithic time, whether by roasting over or in front of the fire, by heating in slices on hot stones, or by wrapping in large leaves and cooking in embers (Tannahill, 1995. In the Neolithic era, however, grains that had been transformed into flour could be prepared in the form of pancakes, galettes, or bread loaves, baked on hot stones (Firmin, 1990). Furthermore, the advent of pottery—the production of waterproof and fire-resistant containers—greatly expanded the possibilities of cooking; it became easy to boil soups, gruels, and purees and to stew all kinds of dishes.

The primary types of Neolithic diet are still very much in evidence in contemporary culinary traditions. For example, the pita bread or the lentil and chickpea purees eaten today by people in the Near East are probably of Neolithic origin. The same applies to the maize tortillas, braised maize cobs, and bean puree of Central America, to the boiled rice of East Asia, to the boiled or cooked-in-embers taro found in Papua, and to the *chuno,* made of frozen, dried, and boiled potatoes in the Andes.

The Propagation of Neolithic Agriculture

Radiating (or expanding) from the Near Eastern, Central American, and East Asian places of emergence, agriculture spread over the millennia to vast regions of the world and gave rise to three major areas of propagation of Neolithic agriculture—the Euro-Afro-Asian, the American, and the East Asian areas, respectively (see Figure 2) (Grigg, 1977). This spread was slow: on the order of one kilometer per year. New vegetal and animal species were domesticated, especially in tropical regions that were rich in easily domesticated plants. These regions became, after the original centers, secondary areas of domestication (see Figure 3) (Gautier, 1990; Harlan, 1992). In each area of propagation, agricultural practices resembled those of the place of emergence from which it was derived. But the secondary areas also offered new characteristics in tools, pottery, or domestic species.

Agriculture of Near Eastern origin, which spread to the Far East in particular, then came into contact with agriculture of East Asian origin. The latter spread to Southeast Asia, where it merged with the New Guinea variety of agriculture, and to South Asia (India), where it came into contact with the Near Eastern form of agriculture. Agriculture of Central American origin merged with agriculture of Andean origin by expanding southward, and with agriculture of North American origin by expanding northward. In this way, maize cultivation was introduced into the middle Mississippi basin by about 300 BCE. Until 900 CE, maize cultivation developed slowly because the maize was of a southern variety and hence ill-adapted to the local region. However, after 800 CE, a variety of maize that was better suited to the cooler local climate appeared. This new variety soon became established and spread north as far as the Great Lakes, so that by about the year 1000 CE, maize, being more adaptable and productive, had largely supplanted the species domesticated earlier in Northeast America, across many parts of this area (Hart & Lovis, 2013).

Regions Unaffected by Expansion

Nonetheless, entire regions of the planet remained unaffected by this first wave of expansion in agriculture as they were too distant from the original places. These regions included Australia, the southern parts of Africa and America, the Northwest of North America, zones of coniferous forest with acid soils unsuited to crops in the far north of America or Eurasia, and, those in the extreme south of America and at the highest altitudes, which proved too difficult to clear with the tools available in the Neolithic period. In Amazonia, Central Africa and Asia, and the great continental plains of North and South America, there was a lack of domestic animals to graze on these regions.

Dissemination of Neolithic Agriculture

The dissemination of Neolithic agriculture beyond its places of emergence occurred mainly through colonization, as the earliest Neolithic cultivators grew in number and expanded beyond their original centers. The descendants of each succeeding generation gradually colonized the areas of propagation—whether unoccupied or occupied by smaller communities of hunter-gatherers, whom they drove away or assimilated (Bellwood, 2004; Cavalli-Sforza, 2001; Renfrew, 1990).

But the propagation of Neolithic agriculture sometimes occurred through the acculturation of hunter-gatherers: the latter, having long lived side by side with Neolithic farmers, borrowed the latter's tools, domesticated their species, imbibed their knowledge and know-how, and thus converted to agriculture. This cooperation partly explains the modifications in tools, habitat, and pottery as agriculture gradually expanded into new territories (Marchand, 2007; Zvelebil & Rowley-Conwy, 1984). Once outside their original centers, Neolithic farmers encountered two types of practically virgin vegetal formations—grassy formations, sometimes with trees or shrubs, and dense, generally closed woodland formations.

The Development of Pastoral Breeding in Grassland and Steppe Zones

Dense grassy formations (temperate grassland and tropical savanna) were difficult to clear with Neolithic tools, and discontinuous grassy formations (cold or arid steppe) were unfit for growing crops; so Neolithic farmers developed the breeding of herbivores there, with or without accompanying crops. The animals in question were goats and sheep on the Mediterranean and Near Eastern higher ground, oxen or zebus in the tropical savannas of Africa and Asia, yaks in the high Himalayan steppes, oxen and horses in the grasslands and steppes of eastern Europe and western Asia, reindeer in the tundra of northern Europe and Asia, dromedaries in the Saharan and Arabian deserts, camels in the deserts of central Asia, or llamas and alpacas on the high Andean grasslands (Grigg, 1977). These breeding activities gave rise to original nomadic or seminomadic pastoral societies, some of which have survived to the present day.

The Development of Slash-and-Burn Agriculture

In contrast, equipped with their polished stone axes, farmers could fell trees in temperate or tropical forests that were not too dense. They then burned the trees and prepared a fertile culture bed by mixing the ashes into the humus on the forest floor. The decomposing humus and the ashes supplied the cultivated plants with the minerals required for their development. In the first cropping season, the yields were high. Thereafter, farmers could still practice supplementary cropping for a few seasons, but fertility was soon exhausted and vegetation spontaneously invaded the plots of land. Farmers then had to let the land lie fallow for several decades and allow the forest and the humus on the forest floor to reestablish before returning to cultivate the same land. It was therefore necessary each year to fell and burn trees on a new land parcel.

Neolithic farmers therefore extended cropping on slash-and-burn terrain to ecologically favorable woody places, characterized by a fairly lengthy growing season and appropriate soils (deep, gently sloping, well-drained, with a reasonably fine texture and a fairly rich chemical composition). This mode of cultivation lasted for millennia and endured up to the present time in the tropical and equatorial forests that still exist in Africa (Cameroon), Asia (mountains of Vietnam, Laos, and Cambodia), and South America (the Amazonian forest of Brazil) (Palm Vosti, Sanchez, & Ericksen, 2005).

The new species domesticated in the areas of propagation led to the emergence of secondary types of diet that were richer and more varied than the primary types. For instance, African millet and sorghum henceforth existed alongside Near Eastern wheat and barley in many regions of Africa; maize and beans from Central America accompanied cassava and groundnuts in many regions of South America; and rice cohabited with sugarcane and taro over vast stretches of East, South, and Southeast Asia.

The Evolution and Differentiation of Agriculture and Food Patterns from the Late-Neolithic Era to the Present

In wooded regions brought under cultivation in the Neolithic period, the population continued to grow over time. This resulted in a greater frequency and intensity of clearing and set in motion a widespread process of deforestation, which, in turn, led to the gradual formation of new ecosystems where slash-and-burn cropping proved increasingly difficult and, ultimately, impossible. In many areas, this resulted in an agro-ecological and food crisis, which lasted until farmers managed to overcome several serious difficulties. Soils that had become too degraded, following deforestation, had to be abandoned or given over to pastoral herding. To continue to cultivate the better land that had lost its wooded fallow, it was necessary to devise new ways to restore fertility. To clear this land—which was now covered with grassy fallow—new tools became necessary. Solving these problems took centuries and sometimes millennia. In the late-Neolithic and early Metal Age, this resulted in new postforest systems of cultivation and/or breeding that were suited to the highly varied ecological conditions in the different regions of the world (Mazoyer & Roudart, 2006).

Systems of Pastoral Breeding and Cultivation in Arid Regions

The wooded savannas covering the Sahara, Arabia, and other poorly watered warm regions in Asia, some ten thousand years ago, were deforested at an early date. These zones were gradually replaced by semi-arid and arid deserts (Cox & Moore, 2010), which, for the last six thousand years or so, have been exploited as poor pastureland by nomadic breeders of sheep, goats, donkeys, and camels, with transhumance from south to north in spring and the reverse movement in autumn.

In the oases and the great valleys irrigated by underground water tables or by rivers with distant sources (such as the Indus, the Nile, the Tigris, or the Euphrates), the Neolithic farmers founded the earliest major civilizations in early antiquity that were based on hydraulic agriculture. Thus, in the valley and the delta of the Nile, which regularly flooded in summer, they developed whole series of basins that were equipped with supply channels and culverts, allowing winter cultivation on receding floodwaters (Hamdan, 1961). In Mesopotamia, which was irregularly flooded by the Tigris and the Euphrates, they built dams, irrigation and drainage channels, as well as protective dykes to combat untimely flooding to practice irrigated cultivation in all seasons (Kramer, 1991).

Systems of Wet Rice Cultivation in the Warm Humid Regions of Asia and Africa

In the deforested regions of monsoon Asia, Neolithic farmers initially grew floating rice (with stalks that extend or retract according to the water level) in natural hollows, which filled with water during the rainy season. They then developed aquatic rice culture in flat-bottomed paddies surrounded by low dykes, where the rainwater filled the paddies and the excess was drawn off by breaches in the dykes. Neolithic farmers in the humid zones of West Africa (the valleys of the Niger, Senegal, Gambia, Casamance, and the Guinea coast) also cultivated various forms of aquatic rice, following the domestication of *Oryza glaberrima*, an African species of rice, in the inner Niger delta and perhaps also in the region of Lake Chad, some 3,500 years ago (Grigg, 1977).

Systems of Hoe Cultivation in the Tropical Savanna

By the end of the Neolithic era, in several average-watered tropical regions in Africa, Asia, or Central and South America, deforestation had led to the formation of grassy savanna or steppe, frequently with trees or shrubs. In Africa, for instance, farmers could develop on the better land a wide variety of hoe-based systems: cereals (millet, sorghum, eleusine) with fallowing and associated pastoral herding (goats, ewes, cattle) in the regions of the Sahel, which received 400–800 millimeters of rainfall; yams in the Sudanese regions (which received 800–1500 millimeters of rainfall); and cropping, without fallowing of teff, Bambara peas, and yams, combined with pastoral herding in the equatorial highlands of East Africa. These cultivation activities grew markedly with the metal hoe, which made it possible to clear dense herbaceous covering after the grass had been burned at the end of the dry season (Fairhead & Leach, 1996).

Systems of Cultivation in European and Mediterranean Temperate Regions

In these regions enjoying average rainfall, deforestation led to the shaping of new, contrasting vegetal landscapes. On the one hand, rough and severely eroded terrain that had become impossible to cultivate constituted the *saltus*, which was given over to pasture for herbivores (ewes, goats, cattle). On the other hand, alluvial plains with a patchwork of cultivated fields formed the *ager*, which produced winter cereals (wheat, spelt, barley, or rye—a belatedly domesticated crop) that alternated as a rule with a short grassy fallow, thereby creating a biennial rotation. This set-aside land was regularly fertilized by the manure of cattle, which were assembled there every evening after grazing all day on the *saltus*. And it was cleared either by deep plowing with spade or hoe, or by semi-plowing with the ard. Enclosed orchards, adjoining the dwellings, formed the *hortus*. And the remaining forestland constituted the *silva* (Burford, 1993). In the course of millennia, these various types of postforest agriculture later became further differentiated.

The Major Agricultural Transformations from Antiquity to the End of the Nineteenth Century

In the monsoon regions of Asia, farmers progressively extended rice cultivation to sloping terrain by constructing terraces to retain the soil and water. Later, principalities, kingdoms, and empires oversaw the construction of large-scale hydraulic installations, which meant that aquatic rice-growing could spread to places where natural conditions rendered it impossible. This involved building terraces on very steep slopes, starting irrigation schemes in insufficiently watered regions, and setting up installations that offered protection from high waters and tides, combined with the regulation of water levels in frequently flooded valleys and deltas (Grigg, 1977).

In the Nile Valley in Egypt, the ancient system of basins and flood-recession cultivation was gradually replaced by one permitting irrigation and cultivation through all the seasons: cotton, rice, or maize in summer; wheat, barley, broad bean, or flax in winter; and sugarcane and various fruits and vegetables all year round (Hamdan, 1961).

In the temperate Mediterranean and European regions, cropping systems of biennial fallowing with associated pastoral herding proved to be unproductive. Throughout antiquity, there was a succession of food shortages, famines, and wars (Garnsey, 1989), so the population remained stable until the tenth century. But after the year 1000, farmers in the northern half of Europe implemented new tools (scythes, carts, plows, harrows, rollers, yokes, shoulder collars) and developed new, more productive cropping and breeding practices (the harvesting and stocking of hay, the stalling of livestock in winter, the production of manure, and the plowing under of organic matter). The resulting new system of cropping and breeding could feed more animals, producing more manure and supporting a far more productive triennial rotation (long fallowing, winter cereal, and short fallowing, spring cereal), which put an end to shortages. The population was better fed and grew considerably (Bloch, 1970; Duby, 1978; Duby, 1990). However, in the fourteenth century, this demographic boom came up against limits that proved insurmountable for a time.

Excessive clearing and a fall in yields triggered a succession of shortages, famines, diseases, revolts, and wars, culminating in the extermination of half the population during the Great Plague of 1347–1350. Once this devastation had been overcome, reconstruction in the Renaissance period of the fifteenth and sixteenth centuries paved the way for the first agricultural revolution of modern times in the northern half of Europe. This revolution was concomitant with the first industrial revolution. It involved replacing fallow land with fodder crops (prairies or root crops) to foster breeding and the production of manure, thereby increasing cereal output. This agricultural revolution, which was started in the sixteenth century by Flemish peasants responding to the growing demand for wool from the nascent cloth-making industry, was subsequently adopted by agricultural entrepreneurs in Britain (during the seventeenth and eighteenth centuries), then by farmers in Northwest Europe (in the eighteenth and nineteenth centuries), and in southern France, northern Italy, and Spain (during the late nineteenth and early twentieth centuries) (Bloch, 1970). These innovations were then transposed to the European settlements in the temperate regions of North and South America, North and southern Africa, and Oceania. Meanwhile, colonial or native plantations of sugarcane, cotton, coffee and cocoa plants, banana and pineapple trees, or tobacco were spread throughout the world to wherever conditions were suitable (Braudel, 1992).

The Contemporary Agricultural Revolution

Having started in the United States in the early twentieth century, the contemporary agricultural revolution had already become widespread in the developed countries by the end of this century, largely supported by public policies (Tracy, 1989; Winders, 2012). It advanced in stages, as industry and research provided more effective mechanical, chemical, and biological resources. Advances included tractors and laborsaving machines of greater power and capacity, yield-enhancing mineral fertilizers for use on crops, concentrated feedstuffs for livestock, and crop-treatment products and veterinary medicines to reduce losses. Selected plant varieties and animal breeds that presented potentially high yields were adapted to these new conditions, making them cost-effective. Meanwhile, new means of transportation, preservation, processing, and distribution meant that farms in different regions could specialize in the products that were the most profitable locally. For example, in Northwest Europe, farm holdings on the continental plains, with cold winters and dry summers, concentrated on the production of cereals and protein-rich oilseeds; farms in coastal regions with milder, damper climates turned toward the breeding of dairy cattle; those at higher altitudes specialized in beef cattle and wood; while farms on warm sunny hillsides favored wine production (Mazoyer & Roudart, 2000). Only a minority of farms have survived all the stages of this process, so more than three-quarters of all the farms that existed in the developed world at the beginning of the twentieth century have now disappeared (Mendras, 1971).

Beginning in the 1960s, a variant of this agricultural revolution, the "Green Revolution," which did not involve the use of powerful tractors and large machines, spread across part of the developing world, especially Asia (India, Pakistan, Indonesia, Philippines, Thailand, etc.). The initial objective of this revolution, largely financed both by public and by private funds from Western countries, was to boost output in response to the food requirements of the population and so check the expansion of communism. The governments of the countries concerned provided resolute support through appropriate policies on prices, credit, research, and agricultural-extension programs. In these countries with industrialization, hundreds of millions of smallholders were able to adopt the new methods, maintain their livelihoods, and increase their incomes (Hazell & Ramasamy, 1991; Mellor, 1998; Trébuil & Hossain, 2004). Conversely, in the countries that lacked the necessary political will, the Green Revolution made only scant progress (Griffon, 2006).

From the 1970s onward, investors of all kinds (entrepreneurs, major landowners, multinational food corporations, various investment funds) extended the agricultural revolution to a number of developing countries. They modernized large properties of thousands of hectares in countries where land and labor were cheap—Latin America (Argentina, Brazil, Mexico), Asia (the Philippines, India), and, to a lesser extent, Africa (South Africa, Zimbabwe). They also created new properties by clearing millions of hectares on the pioneer fronts of vast forests and savannas (Amazonia, Central Africa, Southeast Asia). Finally, since the 1990s, following the collapse of communism, these investors have also become involved in the modernization of former state-owned or collective farms in the former Soviet Union, and in Central and Eastern Europe.

These developments resulted in a spectacular increase in world food production, which grew more from 1950 to 2000 than in all the previous ten thousand years of agricultural history. Output has multiplied by 2.6, making it possible to feed a world population that has been multiplied by 2.4. This increased output has been due mainly to rising yields and only slightly to the extension of the surfaces under cultivation. In countries that have implemented this agricultural revolution, a minority agricultural population produces a sufficient marketable food surplus to feed a very large nonagricultural population (Roudart & Mazoyer, 2012).

However, this agricultural revolution has created a number of roadblocks. In some regions, the overuse of fertilizers and crop treatment products, along with the overconcentration of livestock in enclosed breeding areas, has led to environmental pollution and food contamination. The fact that farms have specialized in a small number of selected species, and even varieties of plants or breeds of animals, has caused a reduction in biodiversity. In some irrigated zones, problems linked to the salinization of soils or groundwater depletion have arisen. On the pioneer fronts, deforestation has advanced. Lastly, these highly productive forms of agriculture are a major source of greenhouse gas emissions (Weis, 2012).

It is nevertheless the case that vast regions of sub-Saharan Africa, Latin America, and Central Asia—along with hundreds of millions of farmers—have remained unaffected by this agricultural revolution. Some two thirds of today's farmers work solely with hand tools, while another third have draft animals; and only a very low percentage work with tractors. Agriculture therefore exists today in many unequal and varied configurations, and all the main types of farming that have existed in the past continue to exist today in one form or another.

Intercontinental Exchanges of Domestic Species and the Diversification in Human Diet

Exchanges between Asia and the Near East-Europe-African Region

Exchanges of domestic species between the Euro-Afro-Asian area of propagation and the Asian area had probably begun as early as the Neolithic era in Central Asia, where farmers originating from each of these areas first came into contact with one another. In antiquity, these exchanges and the acclimatization of plants and animals far away from their places of origin were favored by the great empires that extended from the Eastern Mediterranean to the Indus, such as the Persian Empire and the empire founded by Alexander the Great. However, it was predominantly in the Middle Ages, with the Arab-Moslem empire (seventh through thirteenth centuries), that such exchanges and acclimatization increased in scale (Rostovtzeff, 1986).

The farmers of the Mediterranean rim and of sub-Saharan Africa thus progressively adopted most Asian domesticated plants. More rarely, plants originating in Asia (millet, buckwheat, broad beans, and hemp) were adopted in the cool temperate zones of Europe. Farmers on the grasslands of Europe and the savannas of Africa adopted the horse, native to Central Asia, as early as antiquity. Saharan breeders used the dromedary, which came from Arabia. In the Middle Ages, the buffalo was introduced from South Asia into the Nile Valley and the rice fields of the Po valley, while the Indian zebu appeared in the African savanna. In the other direction, most of the plants and animals domesticated in the Near Eastern place were adopted into Northeast Asia. And most of the plants domesticated in sub-Saharan Africa appeared in South and Southeast Asia.

Exchanges between the Old and the New Worlds

It is indeed possible that exchanges of domestic species between the Old and the New Worlds took place before the period of the great maritime explorations, with calabashes or cotton simply floating across the oceans or being transported by sailors between continents (Harlan, 1992). But it was really from the fifteenth century onward, after the discovery of America by the Europeans, that large numbers of plants that had been domesticated in the Americas began to circulate in Europe. Thereafter, these species were spread more widely in subtropical Africa and Asia, along with many other tropical species. In the opposite direction, the lentil and pea from the Near East, the rice and soya from Asia, and domestic animals such as the cow, the pig, the sheep, or the horse (which were indigenous to the Old World), greatly contributed to the transformation of the various forms of agriculture on the American continent, where only a few animal species had been domesticated (Grigg, 1977).

Enrichment and Diversification of Human Diets

These exchanges of domestic species over vast distances broadened the range of cultivated or bred species, which entailed further enrichment and diversification in human diets. Thus, many national dishes or local specialties contain both species that have been domesticated in a given region in Neolithic times and those that have been brought from elsewhere at a later stage. Brazilian *Feijoada*, for instance, includes cassava and beans from America along with rice from Asia and pork from the Near East. The hamburger with French fries of North America is a combination of wheat bread and beef from the Near East, with the potato and the tomato coming from America. One very popular variation on the Indian *dhal* brings together Near Eastern lentils and American tomatoes, which is served with Asian rice. The *Mafé* of Senegal, which accompanies many dishes throughout West Africa, is a sauce made of peanuts, which were originally American. A number of symbolic specialties—such as paella in Spain, made of rice, tomatoes, and peppers, or the caprese salad in Italy, which includes tomatoes, mozzarella from buffalo milk, and capers from Iran, are entirely made up of ingredients that come from afar—in these particular cases, from America and Asia.

Conclusions and Observations

For ten thousand years, the provision of food for the human population has been determined by the development of agriculture, which, in turn, has been determined by the ecological characteristics and the technical and cultural circumstances prevailing in each place at a particular time. Agriculture emerged in the Neolithic era, in a few small ecologically and culturally privileged areas. In these places of emergence, human food production was concentrated on domesticated species. From these original centers, Neolithic farmers propagated their practices to suitable and much more extensive regions, with slash-and-burn cropping both in tropical and in temperate wooded regions and the pastoral herding of herbivores in grassland regions. In each of the areas to which agriculture spread, farmers domesticated new species, thereby diversifying their food supply. Over time, with rising populations, slash-and-burn activities were intensified, eventually leading to deforestation, new ecosystems, and new forms of agriculture and food production.

In the course of recorded history, the process whereby forms of agriculture and diet have become diversified has continued. Great travelers carried species, domesticated in Neolithic times, across the continents, where farmers would usually attempt to acclimatize them. Wherever these attempts were successful, new foods were adopted, often without displacing those that already existed, like so many historical layers of sediment. In some regions, production techniques greatly improved, especially during the nineteenth and twentieth centuries, so that modern output has greatly outpaced what farmers themselves consume. This surplus can be transported to distant markets and consumed by the ever-growing, nonfarming urban populations, sometimes in cities situated thousands of kilometers away and inhabited by tens of millions of people. Yet, even today, in each area to which Neolithic agriculture had once spread, the diet of most people is still largely based on the selfsame species as it was in the late-Neolithic era. Thus, Americans consume more corn and beans than do the inhabitants of other continents, Europeans eat more bread cereals and peas, Asians more rice and soya, Africans more millet, sorghum, yams, and Bambara peas, and Papuans more taro (Collomb, 1999).

Admittedly, most people can obtain food products from all over the world. And a few convenient staples (rice, pasta, bread) or fast foods (pizza, hamburgers, French fries, sushi) are eaten throughout the world. One might therefore conclude that a single universal diet is gradually emerging. But each of these internationalized products is consumed by only a small part of the population, often only on an occasional basis, which does not make it a single universal diet. It appears that the culturally diverse diets found in different areas of the world are still developing and that this development is based on locally produced staple foodstuffs.

References

Bellwood, P. (2004). *First farmers: The origins of agricultural societies.* Malden, MA: Wiley-Blackwell.

Bloch, M. (1970). *French rural history: An essay on its basic characteristics.* Berkeley: University of California Press.

Bocquet-Appel, J.-P., & Bar-Yosef, O. (2008). *The Neolithic demographic transition and its consequences.* Dordrecht: Springer.

Braudel, F. (1992). *Civilization and capitalism, 15th–18th century, vol. I: The structures of everyday life.* Berkeley: University of California Press.

Bulmer, S. (1975). Settlement and economy in prehistoric Papua New Guinea: A review of the archeological evidence. *Journal de la Société des océanistes, 31*(46), 7–75.

Burford, A. (1993). *Land and labor in the Greek world.* Baltimore, MD: Johns Hopkins University Press.

Cauvin, J. (2007). *The birth of the gods and the origins of agriculture.* Cambridge: Cambridge University Press.

Cavalli-Sforza, L. L. (2001). *Genes, peoples, and languages.* Berkeley: University of California Press.

Collomb, P. (1999). *Une voie étroite pour la sécurité alimentaire d'ici à 2050.* Paris: Economica; FAO.

Cox, C. B., & Moore, P. D. (2010). *Biogeography: An ecological and evolutionary approach* (8th ed.). Hoboken, NJ: Wiley.

Demoule, J.-P. (Ed.). (2009a). *La révolution néolithique dans le monde.* Paris: CNRS Editions.

Demoule, J.-P. (2009b). Naissance des inégalités et prémisses de l'Etat. In J.-P. Demoule (Ed.), *La révolution néolithique dans le monde* (pp. 411–426). Paris: CNRS Editions.

Duby, G. (1978). *The early growth of the European economy: Warriors and peasants from the seventh to the twelfth century.* Ithaca, NY: Cornell University Press.

Duby, G. (1990). *Rural economy and country life in the medieval West.* Columbia: University of South Carolina.

Fairhead, J., & Leach, M. (1996). *Misreading the African landscape: Society and ecology in a forest-savanna mosaic.* Cambridge: Cambridge University Press.

Firmin, G. (1990). Le passage d'une économie de prédation vers une économie de production au néolithique en France. In C. Chadefaud, G. Firmin, & J. Villemonteix (Eds.), *Agriculture, plantes utiles, alimentation, cuisine chez les néolithiques, les Egyptiens et Grecs anciens.* Poitiers: Faculty of Sciences, University of Poitiers.

Food and Agriculture Organization of the United Nations) (FAO) (2014). FAOSTAT Food balance. Retrieved from http://faostat3.fao.org/browse/FB/*/E.

Garnsey, P. (1989). *Famine and food supply in the Graeco-Roman world: Responses to risk and crisis.* Cambridge: Cambridge University Press.

Gautier, A. (1990). *La Domestication: Et l'homme créa ses animaux.* Paris: Editions Errance.

Griffon, M. (2006). *Nourrir la planète.* Paris: Odile Jacob.

Grigg, D. B. (1977). *The agricultural systems of the world.* Cambridge: Cambridge University Press.

Guilaine, J. (Ed.). (1991). *Prehistory: The world of early man*. New York, NY: Facts on File.

Hamdan, G. (1961). Evolution of irrigation agriculture in Egypt. In L. D. Stamp (Ed.), *A History of land use in arid regions* (pp. 119–142). Paris: UNESCO.

Harlan, J. R. (1992). *Crops and man* (2nd ed.). Madison, WI: American Society of Agronomy-Crop Science Society.

Hart, J. P., & Lovis, W. A. (2013). Reevaluating what we know about the histories of maize in northeastern North America: A review of current evidence. *Journal of Archaeological Research*, *21*(2), 175–216.

Hazell, P. B. R., & Ramasamy, C. (1991). *The green revolution reconsidered: The impact of high-yielding rice varieties in South India*. Baltimore, MD: Johns Hopkins University Press.

Jiang, L., & Liu, L. (2006). New evidence for the origins of sedentism and rice domestication in the Lower Yangzi River, China. *Antiquity*, *80*(308), 355–361.

Kramer, S. N. (1991). *History begins at Sumer: Thirty-nine firsts in man's recorded history* (3rd ed.). Philadelphia: University of Pennsylvania Press.

Liu, L. (2007). *The Chinese Neolithic: Trajectories to early states*. Cambridge: Cambridge University Press.

Marchand, G. (2007). Neolithic fragrances: Mesolithic-Neolithic interactions in western France. In *Proceedings of the British Academy* (Vol. 144, pp. 225–242). Oxford: Oxford University Press.

Mazoyer, M., & Roudart, L. (2000). The socio-economic impact of agricultural modernization. In *The state of food and agriculture* (pp. 171–198). Rome: FAO.

Mazoyer, M., & Roudart, L. (2006). *A history of world agriculture: From the Neolithic age to the current crisis*. New York, NY: Monthly Review Press.

Mellor, J. W. (1998). Foreign aid and agriculture-led development. In C. K. Eicher & J. M. Staatz (Eds.), *International agricultural development* (3rd ed., pp. 55–66). Baltimore, MD: Johns Hopkins University Press.

Mendras, H. (1971). *The vanishing peasant: Innovation and change in French agriculture*. Cambridge, MA: The MIT Press.

Palm, C. A., Vosti, S. A., Sanchez, P. A., & Ericksen, P. J. (Eds.). (2005). *Slash-and-burn agriculture: The search for alternatives*. New York, NY: Columbia University Press.

Pernès, J. (1983). La génétique de la domestication des céréales. *La Recherche*, *14*(146), 910–919.

Renfrew, C. (1990). *Archaeology and language: The puzzle of Indo-European origins*. New York, NY: Cambridge University Press.

Rollefson, G. O., & Köhler-Rollefson, I. (1992). Early neolithic exploitation patterns in the Levant: Cultural impact on the environment. *Population and Environment*, *13*(4), 243–254.

Rostovtzeff, M. I. (1986). *The social and economic history of the Hellenistic world*. Oxford: Oxford University Press.

Roudart, L., & Mazoyer, M. (2012). Origins, development and differentiation of world agricultures. In M. Agnoletti, E. Johann, & S. Neri Semeri (Eds.), *Encyclopaedia of Life Support Systems—World Environmental History*. Oxford: UNESCO–EOLSS Publishers.

Sahlins, M. (1974). *Stone Age economics*. Chicago, IL: Aldine Transaction.

Skorupa, H. (2013). *Food preparation methods used in the Neolithic Period*. Norderstedt, Germany: GRIN Verlag.

Smith, B. D. (2007). *Rivers of change: Essays on early agriculture in Eastern North America*. Tuscaloosa: University of Alabama Press.

Tannahill, R. (1995). *Food in history* (2nd ed.). New York, NY: Broadway Books.

Testart, A. (2012). *Avant l'histoire: L'évolution des sociétés, de Lascaux à Carnac*. Paris: Gallimard.

Tracy, M. (1989). *Government and agriculture in Western Europe* (3rd ed.). New York, NY: Prentice Hall/Harvester Wheatsheaf.

Trébuil, G., & Hossain, M. (2004). *Le riz: Enjeux écologiques et économiques*. Paris: Belin.

Weis, T. (2012). *The global food economy: The battle for the future of farming*. London: Zed Books.

Winders, B. (2012). *The politics of food supply: U.S. agricultural policy in the world economy*. New Haven, CT: Yale University Press.

Zvelebil, M., & Rowley-Conwy, P. (1984). Transition to farming in Northern Europe: A hunter-gatherer perspective. *Norwegian Archaeological Review*, *17*(2), 104–128.

Chapter 2

Agricultural Industrialization and the Presence of the "Local" in the Global Food World

Carlos J. Maya-Ambía

Introduction

Since the Neolithic Era, agriculture has been the main economic activity in most societies, and, more importantly, this activity remained essentially unchanged until the nineteenth century, when the Industrial Revolution emerged (Higman, 2012; Standage, 2009). Agriculture became an unavoidable source of raw materials for industrial processes, and rural populations began migrating to cities, where industries and new opportunities were developing. At the same time, an abundant supply of cheap food and ingredients for food production contributed significantly to keeping urban wages low enough to warrant high and continuous profits to the capital invested in industrial activities. These developments established a geographical division of labor between the cities and their rural surroundings. With some exceptions, such as "exotic" and luxury goods for the upper urban social classes, the bulk of food was produced in rural areas surrounding the cities. This meant that the places of food production and food consumption were close, not just geographically, but culturally. As a consequence, there were strong differences in food patterns between places around the world. This scenario changed dramatically due to globalization, namely since the last quarter of the nineteenth century, but it was particularly accelerated by the industrialization of agriculture due to the Green Revolution following World War II (Pilcher, 2006).

Regarding agriculture and food, I agree with McMichael (2009a, 2009b) and Burch and Lawrence (2009), among others, who speak of three food regimes in the globalization process, but it is also important to mention that other scholars suggest we are currently in transition between the second and third food regimes (Friedman, 2009). *Food regime* is a historical concept that "has demarcated stable periodic arrangements in the production and circulation of food on a world scale, associated with various forms of hegemony in the world economy: British, American, and corporate/neoliberal" (McMichael, 2009a, p. 281). According to Antonio Gramsci, *hegemony* means economic and political dominance as well as intellectual and moral guidance, and the concept is employed in this sense here (Buci-Glucksmann, 1980). In other words, a food regime is a period of stable political and economic relationships operating on a global scale. It is not merely about food; but focusing on food, one can understand the dynamic of the global system and the tensions between control and subordination, between dominant and dominated social actors, between rich and poor regions and places (Friedmann, 1982; Friedmann & McMichael, 1989; McMichael, 2009a, 2009b).

The first food regime existed approximately from 1875 to the start of World War I. It was centered on the hegemony (political dominance) of Great Britain and the expansion of British industry and finance capital. At the same time, the international trade flows of food were dominated by wheat. Wheat was mainly produced in Argentina, Australia, and the United States, and it was exported to Great Britain and other industrial centers. This food regime experienced a crisis, together with the breakdown of nineteenth-century civilization, as discussed in detail by Karl Polanyi (1944). After a long period of economic and political turbulence, a second food regime emerged around 1945, under the hegemony of the United States. It functioned for approximately three decades and was characterized by a new division of labor. Enormous agricultural surpluses from the United States flooded the rest of the world, particularly in those countries that needed help, such as postwar Europe and, later, the Third World, including countries within Latin America, Asia, and Africa. This "Food Aid" from the United States was used as an effective weapon against revolutionary movements in underdeveloped countries and as a barrier to the expansion of socialism during the Cold War (Friedmann, 1993; Hobsbawm, 1994; Magdoff & Tokar, 2010; Pilcher, 2006). During this period, Washington promoted the expansion of American big enterprise on a global scale, and, specifically, Washington imposed the American way of agriculture. The Green Revolution became the model for agricultural practices around the world. This model featured the use of "improved seeds," and the systematic application of chemical fertilizers, pesticides, and herbicides (Magdoff, Foster, & Buttel, 2000; Sale, 1993; Segrelles, 2001; Shiva, 1992).

By the mid 1970s, a remarkable crisis took place in the American food regime, and the third food regime emerged. This regime was powered by multinational corporations, which are present at every stage of food production. At the level of international trade, fresh fruits and vegetables became the most important items being produced. According to Burch and Lawrence (2009, p. 275), during the third food regime, "many developing countries have been incorporated into the supply chain as sources of cheap processed foodstuffs and of fresh fruit and vegetables demanded by consumers in the North." On the other hand, the local division of labor presents certain features. The sites of multinational corporations are mainly cities in the United States, Western Europe, and Japan, where the populations are the most conspicuous importers of fresh fruits and vegetables produced in underdeveloped countries. Although it does not seem that the third food regime has been exhausted, it has undoubtedly contributed to the current ecological and food crises, described later in the chapter. At this point, it seems appropriate to provide some context for the precise use of the concept(s) of *local* and *place* in this chapter.

Alfred Weber, and similar authors and economists, use an analytical framework of "industrial location," and the World Bank employs the concept of "local economic development"; but I propose a wider meaning of the word "local," following the tradition of the Mexican historian Luis González (1974), founder of *microhistory* as a social science. In his seminal work, the "local" refers to a small and relatively isolated village, San José de Gracia, where nothing extraordinary occurs, and daily life is shaped by traditions, religion, routine economic activities, political happenings, and some external influences. The village, whose history is studied by González, has been used as a model of place: of any place of the world, understood as *matria* (motherland), in contrast with *patria* (fatherland): the native land in contrast with the big country, the homeland versus the metropolis. This kind of model is particularly useful in the study of food, because eating is basically a routine activity, learned in childhood and repeated throughout adulthood. At the same time, the production, processing, marketing, and consumption of food are influenced by a constellation of diverse factors: geographical, economical, political, historical, social, religious, etcetera. Accordingly, a *place* is more than its hectares, and even more than a political administrative unit; it is the intersection of several realms of political institutions, human activities, beliefs, habits, desires, experiences, and memories. The geographical extension of a place depends upon its homogeneity, its common history, its shared ethical and aesthetic values, and its self-image as a unity.

Agricultural Practices: Beyond the Limits of Production-Consumption

Traditionally, agriculture has been considered an important sector of the national economy or even as an industry focused on the production of certain goods. As in other industries, agriculture has been studied from the production of goods (the first stage) until the consumption of these goods (the final stage). Given the complexities of globalization, this approach is very limited and does little to help us understand real agricultural trends. On the contrary, the scenario becomes clearer if we consider agriculture as a global system (Friedland, 2001) and as a long global value chain, composed of several links where agents interact and connect with the whole economy, nationally and globally (Gereffi & Korzeniewicz, 1994). Accordingly, the global economy is formed by a complex web of value chains, whose links are located in different places around the world. Therefore, it is correct to speak of global value chains, especially the global value chain of agriculture that does not begin at the production process, but rather with the appropriation of nature and the transformation of natural objects into economic inputs, including the current land-grabbing in several places by transnational corporations (Belasco & Horowitz, 2009). Driven by profit, these corporations have appropriated land, resulting in disastrous ecological effects (Shiva, 2013). Also, this value chain does not end at the consumption phase, but continues with the negative consequences of consumption—such as environmental hazards that damage soil and groundwater. These practices of appropriation and consumption have created a "new international division of labor": the Global South has become the place of appropriation of nature and in some ways a type of dumping ground. Yet, according to the World Trade Organization (WTO, 2012, Table 11.20), the leading food importers in the world are: the European Union (EU) (36.4 percent of world imports in 2012), the United States (8 percent), China (6.2 percent), Japan (5.4 percent), and Russia (2.7 percent). Among the most relevant food exporters are developed countries, such as EU members, the United States, Canada, Australia, and Norway, and also a large group of developing countries, including Brazil, Chile, Indonesia, Argentina, India, Thailand,

Malaysia, Mexico, and Vietnam. In the case of developed countries, their food exports are usually less relevant than their manufactured exports. On the contrary, for many poor countries, food exports represent more than 30 percent of their total exports, and they are are strongly dependent on these types of goods. This is the case in countries such as Ethiopia, Malawi, Nicaragua, Uruguay, Paraguay, Argentina, Uganda, Guatemala, Ghana, Kenya, Côte d'Ivoire, Honduras, Costa Rica, and Brazil (WTO, 2012, Table 11.21). At the same time, these places in the Global South are usually the cultivation zones of high-value agricultural products (fruits and vegetables) consumed by the high-income populations located in the Global North, where the most food waste is produced (Brown, 2012; Goodman & Watts, 2013). Certain agents dictate the chain's direction and impose the standards that participants must observe, driving the global value chain of agriculture. It is important to ascertain who exerts this governance of the value chain and, more significantly, which of its links dictate the entire dynamic. In the remaining sections of this chapter we will examine each link of the agricultural global value chain in an attempt to elucidate who exerts the governance and how these powers pertain to local conditions or place.

Transnational Corporate Control of Seeds Production

Until recent times it has been traditional for farmers to keep some seeds from their yearly crops to use in the next planting cycle. This practice has been radically modified by the introduction of the so-called improved or hybrid seeds, and recently the genetically modified seeds, which are produced by a handful of big multinational enterprises such as Monsanto (Shiva, 2000). The farmers who work with these seeds must buy them every year, because corporations use patent law to protect the reuse of a licensed, patented seed. According to the ETC Group (2008), the top ten companies account for 67 percent of the global proprietary seed market. Monsanto is the world's largest, and its market share is 23 percent. The American company DuPont has the second place (15 percent of the world market), followed by Syngenta (Switzerland), Limagrain (France), Land O'Lakes (United States), KWS AG (Germany), Bayer (Germany), Sakata (Japan), DLF-Trifolium (Denmark), and Takii (Japan).

The transnational corporate control of seeds production is even stronger from the perspective of the farmers due to the expansion of genetically modified organisms (GMO) and the claims of transnational corporations on property rights. Originally, these seeds came from the Global South, but now companies located in the Global North control them. The important role of place becomes particularly clear when we consider that there are a number of countries in the Global North where the use of GMO seeds is prohibited, while in the Global South their use is encouraged. It is worth noting that the activities of multinational corporations such as Monsanto can be limited by political forces and institutions whose power and effectiveness vary according to conditions of place. For instance, after strong worldwide protests against Monsanto, the company decided to withdraw from the EU except for Spain, Portugal and the Czech Republic. The reason was that these countries, unlike Germany and France, had no social movements defending the environment. Also, the strength of democratic institutions, political parties, parliaments, and workers' unions clearly plays an important role in decision-making. Another example is India, where there is a vibrant social movement, inspired by Vandana Shiva (2007), promoting the practice of saving and exchanging seeds among the farmers. But these are exceptions; companies such as Monsanto dictate the rules of the game to farmers around the world. Most world farmers, especially those in underdeveloped countries, depend on seeds produced by multinational companies.

At this point it is very important to recognize the levels of organization of farmers and the agrarian policies of each country (OECD, 2007). Both components differ from country to country, and it is well known that the strongest farmers' organizations are in the EU. This explains, in part, why Monsanto has found strong resistance in Western Europe.

Chemical Inputs of Industrialized Agriculture

In contrast to the traditional seeds used for millennia, genetically engineered seeds do not work properly without the massive employ of agrochemical inputs, such as herbicides, pesticides, and fertilizers. The farmers involved, willingly or unwillingly, depend on the supply of these items, meaning that they must accept the prices fixed by the corresponding companies. In many cases the same company produces seeds as well as corresponding inputs. The biggest agrochemical firms are Syngenta, Bayer, BASF, Dow AgroSciences, Monsanto, DuPont, Makhteshim Agan, Nufarm, Sumitomo Chemical, and FMC. One must keep in mind that the world's six largest agrochemical manufacturers, who control nearly 75 percent of the global pesticide market, are also seed-industry giants. Their increasing profits provide a glimpse of only one side of the coin. The other side shows that agrochemicals used in industrialized agriculture have caused severe damage to the environment and to people—both to those who are in direct contact with them and to many others living near the fields. Neurological diseases, birth defects, infertility, stillbirths, miscarriages, several kinds of cancers, and many other complications have been caused by frequent exposure to agrochemicals (Schlanger, 2014; WHO, 2014a; 2014b). Though the majority of exposure victims are people living in the Global South (Eddleston et al., 2002), these substances are also affecting the lives of an increasing number of US citizens (Ongley, 1996).

One example of this situation is in farming communities in Central America, Sri Lanka, and India, where most people with kidney disease are agricultural, farm, or sugarcane workers. Researchers have pointed out the probable relationship between kidney disease and significant amounts of arsenic and heavy metals found in fertilizers and pesticides that are available to the farmers in the affected regions (Yaqub, 2014). A second example is hypothyroidism in India, where environmental factors such as unregulated pesticide use might play a part in this disease (Bagcchi, 2014). Another example is Mexico, where fourteen years ago, it was already detected that this country

> is the agricultural zone with the highest health damage to its population from pesticides. (And) a strong relation was found between the health problems and the pesticides applied. Finally, the study concludes that several Mexican States appear as a risk zone because of contamination by pesticides that are permitted by the regulation norms and by others whose use is prohibited in other countries and in Mexico. This situation causes a serious effect on the health deterioration of people who are exposed to these agro-chemicals and, in some cases, leads to death. (Valdez, García, & Wiener, 2000, p. 399)

Unfortunately, more recent studies show that health damage persists as the result of agrochemical use (Beraud, Galindo & Covantes, 2008; Carías, 2014; Moreno & Lopez, 2005; Riojas, Schilmann, López, & Finkelman, 2013).

It appears that one of the primary motivations of the multinational agrochemical companies is short-term profit. Damage to the environment or to people is not considered unless strong political and social institutions and movements force the companies to take such problems into account. In this case the role of national or even regional governments and parliaments is relevant, because they can legally regulate the activities of the mentioned companies. This explains why many of these products cannot be used in developed countries, even as they are being employed on a massive scale in underdeveloped regions. This uneven exposure is related to the existence or nonexistence of democratic and honest governments, who effectively impose regulations, as well as to the strength and responsiveness of the social movements in each place. For example, five years ago, researchers from the University of Guadalajara discovered high concentrations of heavy metals and five kinds of pesticides in the river Santiago, which is close to the city of Guadalajara (*El Universal*, Guadalajara, July 9, 2009). Each of these metals and chemicals is prohibited in the United States. There are more than thirty such pesticides that are prohibited in other countries but are allowed in Mexico. Between 2001 and 2003, twenty-seven million pounds of pesticides were imported from the United States, and they are still being exported from the United States to Mexico (Smith, Kerr, & Sadripour, 2008). Under different political conditions, in the presence of a strong social movement, this type of wanton environmental degradation would be less likely.

By this way, seeds and agrochemicals combine to produce agricultural goods. Most of them are channeled to domestic markets, but an increasing share enters into foreign markets—hence the relevance of international trade flows of food items.

The Control of International Trade Flows

The beginning of this chapter noted that one relevant feature of the current third food regime is the international trade of fresh fruits and vegetables (Arce & Marsden, 1993; Dolan & Humphrey, 2004; McMichael, 2009b; Kritzinger, Barrientos, & Rossouw, 2004). Actually, international trade in fruits and vegetables has expanded rapidly since the end of the 1970s, when "the average value share of fruits and vegetables (including pulses and tree nuts) in global agricultural exports increased from 11.7 percent in the period 1977–81 to 15.1 percent in 1987–91 and reached an all time high of 16.5 percent in 1997–2001" (Huang, 2004, p. 3). This trend has continued, and in 2010 fruit and vegetables constituted the most important group in the composition of global agricultural exports (FAO, 2013, p. 151).

At the same time, it should be noted that globalization is understood mainly as unlimited flows of investments and trade. Under these conditions, the trade of agricultural products is strongly concentrated in a handful of transnational companies. The trade policies of such companies affect (positively or negatively) the local development of places (regions or nations) taking part in the international agricultural trade. This is particularly relevant in the case of food products, and it became evident during the recent "food crisis" (Magdoff & Tokar, 2010). A special aspect of the international food trade is the role played by transnational supermarket chains. These chains follow strategies that affect the economic, social, and health circumstances of populations around the world. Each social group reacts to these problems, but its effectiveness varies according to local conditions—namely economic and political institutions—in addition to cultural traditions (Dolan & Humphrey, 2004; Gereffi & Christian, 2009; Humphrey, 2007; Watson & Caldwell, 2005).

The expanding presence of supermarket chains around the world is affecting both producers and consumers. One likely effect on agricultural producers is the worldwide expansion of competition. Local suppliers are now in permanent competition not only with one another, but also with farmers located in other countries and on other continents. A kind of homogenization takes place on the supermarket shelves, where offered items and their differences are reduced to quality (color, taste, freshness) and price. The social, political, and economic reality behind quality and price is completely hidden. Consumers often know nothing about the production of the food they are buying. They likely know nothing about the living conditions of the labor force employed by the agricultural companies: if children are hired; if workers, including children and women, are laboring under safe conditions; if they are already suffering the deadly effects of chemical substances; if they must be transported thousands of kilometers from their born places under inhumane circumstances; if they are women who must leave their babies in the care of other people; or if they are men who must take their families with them to work on the fields. The consumer is also often misinformed about the political structures and institutions behind the prices—for instance, subsidies, compensation payments, and export supports. The agricultural policies followed by different governments, vary widely (Morgan, Marsden, & Murdoch, 2006). Therefore, the supermarket shelves are filled with competing goods that are not equal—with uneven costs and consequences across place and population—yet they appear equal to the consumer.

The principal consequences of competition between farmers are impacts to local and regional development, mainly in zones that are firmly oriented toward international agricultural markets. Only the most capable local farmers can become suppliers to supermarket chains, which increases the gap between the poor and the rich (or less poor) local farmers (Humphrey, 2007). On the one hand, local farmers can get more money for their produce from powerful supermarket chains like Walmart, and thus participate in the global economy. Yet, on other hand, they are also subject to stricter standards and regulations that many farmers do not have the capital to meet (Dolan & Humprey, 2004). Supermarket companies do not always pay immediately; the farmers must sometimes wait thirty, sixty, and even ninety days to get their money. In other words, farmers are financing, at least in part, the activities of supermarkets.

Supermarket chains have other global activities that impact consumers. If consumers have no other options, such as local green groceries, farmer markets, organic markets, etcetera, they must accept the products offered by the supermarket chain. These limited choices often impact the traditional, culturally evolved diets of diverse societies. Strong campaigns through television, newspapers, and social media are promoting these changes, which can be positive or negative for the consumer's health. The outcomes remain controversial. However, these local consumption traditions must be taken into account by the supermarket chains. They can introduce new food products, but if they want to be successful they must also offer traditional local foods (Meyer & Bernier, 2010; Staertzel, n.d.; Watson & Caldwell, 2005). It is also important to consider special interests of the consumers. For instance, in poor countries, price is perhaps the most important deciding factor for consumers buying food. But in developed nations such as Japan (Shirai, 2010), appearance, presentation, freshness, and convenience can be more important than price. Moreover, it must be taken into account that in developed countries, many consumers are interested in "fair trade" and other socially conscious purchases. Consumers hope their items have been produced under fair social and environmental conditions (Fairtrade Canada, 2012; Ferran & Grunert, 2007). Something similar is happening with organic foods, which helps explain why multinational supermarket chains are introducing these kinds of products on their shelves (Canavari & Olson, 2007; Willer, Yussefi-Menzler, & Sorensen, 2008). The internationalization of food occurs through international trade of food items as well as through the food service sector. This allows unheard of combinations of global and local dimensions of the worlds of food (Morgan, Marsden, & Murdoch, 2006).

Transnational Food-Service Companies and the Local Consumption of Global Foods

The food service sector is controlled by well-known transnational companies, which undoubtedly influence global consumption patterns. Some undesirable health consequences of their products include malnutrition and obesity. But at the same time, these companies must adapt to the food culture and traditions of the places where they operate (Gottdiener, 2000; Watson & Caldwell, 2005). If they are unable to do this, they may fail and eventually exit the market. Of course we are referring to multinational food and beverages companies such as McDonald's, Burger King, Pizza Hut, Domino's Pizza, Coca-Cola Company, PepsiCo, Ferrero, General Mills, Grupo Bimbo, Kellogg's, Kraft Foods, Mars, Nestlé, Unilever, and Taco Bell. Their products include baby food, bakery items, canned/preserved food, chilled/processed food, dairy, dried/processed food, frozen items, ice cream, meal replacements, noodles, oils and fats, pasta, ready meals, sauces, dressings and condiments, snack bars, soup, spreads, snacks, and soft drinks.

Several factors have contributed to the success of these fast-food suppliers (Wilk, 2006). The first factor is labor conditions in big cities, where workers—many of whom are women—must get up very early in the morning and spend long hours commuting by car, bus, or train to their workplaces. Two-way commutes leave these workers with little time for buying, preparing, and cooking their own meals. Under such conditions, fast food becomes a common alternative. A second factor contributing to fast-food-supplier success is the increasing number of single households—young people who no longer live with their parents, who do not want to marry, or who have limited marriage opportunities. For these consumers, fast food is cheaper and/or more convenient than cooking in the home. A third factor is the massive and pervasive advertising promoting the consumption of these products, which sooner or later influences consumers around the world. In spite of these advantageous factors, the multinational companies must still accept some kind of compromise between their global interests and the local consumption patterns of the places where they operate. The following are examples that help illustrate this quandary.

In Japan, where I spent almost a year across several stays between 2006 and 2014, I observed that Pizza Hut offers the most common kind of pizzas one can find around the world, mainly those made with tomato sauce, ham, and basil. At the same time, the menu includes pizzas that are probably only consumed in Japan—for instance, pizzas prepared with squid, different types of seaweed, or *nira* (a type of Japanese leek). The menu combines pasta with seaweed, *kinoko* (a Japanese mushroom), teriyaki chicken, or grilled pork meat (*Kagoshima* style). Furthermore, the menu offers the Japanese consumer dishes that are full of Mexican jalapeño chilies and avocado. Together, the dishes demonstrate a blend of global and foreign ingredients among the traditional Japanese food items. This is also a notable feature of Japan's dietary transition (Francks, 2009; Smil & Kobayashi, 2012).

A second example is Taco Bell and its introduction into Mexican culture. The company headquartered in Irvine, California, tried to teach the Mexicans, the creators of *tacos*, how to consume them. Taco Bell brought to the Mexican consumers *tacos* made of ground beef and yellow cheese, but most consumers found this taste impossible to accept. After just one year, the company withdrew from Mexico (Gutmann, 2000; Stevenson, 2007).

The Uneven Distribution of Benefits and Undesired Effects

The global value chains of agricultural and particularly of food products present two primary characteristics. The first one is the uneven distribution of benefits among the participants in those chains, where the smallest share goes to the direct producers—the farmers—who, in different countries, including those of the developed world, receive the smallest share, particularly if they have small farms (Bailey, 2011; Magdoff, Foster & Buttel, 2000; Magdoff & Tokar, 2010). This can be observed by comparing the long-run trends of producer prices against the consumer prices of the most important food items. The former tend to stagnate while the latter show an increase. The same trend has been observed in the EU, where farmers labor under better conditions than those in other countries (European Parliament, 2007; Pretolani, Cavicchioli, & Cairo, 2014); in the United States (Lee, Summer, & Vergati, 2012; Schnepf, 2013); and in numerous developing countries (CEPAL/FAO/IICA, 2011).

There are several explanations for this gap. The main causes can be expressed by one simple concept: governance. Factors such as market power, financial and technological capabilities, leadership, and even cultural influence constitute governance. The agents who exert their governance in the value chain receive the greatest share of total value produced in the chain. To envision this structure of power, imagine an hourglass shape. There are thousands of direct producers at the top of the hourglass, and there are thousands of end consumers at the bottom. They are connected by a small number of companies located in the middle. Under current conditions, this middle section represents the supermarket chains, but other agents, such as seed companies, are represented there too. This is due to the oligopolistic structure of the market, where a few companies can collude and impose product prices that must be accepted by farmers who have no other choice. Agrochemical markets may also exist in this middle section, because some multinational enterprises produce seeds, fertilizers, herbicides, and other important inputs for industrialized agriculture. This vertical integration reinforces their governance in the food-value chain and allows them to manipulate the prices of their products. Some companies have extended their influence further by branching out into various economic areas—for example, supermarkets offering banking operations.

The second factor worth mentioning is the skewed distribution of the harmful effects of industrialized agriculture. The first phase is the destructive appropriation of nature—for instance, the violent transformation of jungles and forests to croplands and pastures. A tragic, current example can be found in the Amazon jungle, where every day nearly seven thousand hectares are destroyed. But the continued loss of forests and woodlands across the globe, including the loss in developing countries, has been an ongoing phenomenon for centuries. The next harmful step is the intensive use of these areas as monoculture lands, which results in depletion and impoverishment of soil and in a reduction of biodiversity. Next, excessive use of chemicals often contaminates the groundwater, and the main victims of the inappropriate use of these substances are human beings. More often the victims are the farm workers and their families, who are in direct contact with the chemicals. The regrettable outcomes include cancer, leprosy, birth defects, and respiratory diseases that mainly affect the poor of developing countries (WHO, 2014a; WHO, 2014b). It is clear that where one lives is relevant. It is not the same thing to live in a middle- or high-class neighborhood in a European or North American city as it is to dwell in a poor village in India, Latin America, or Africa.

Another important problem related to food consumption is malnutrition. Malnutrition includes undernutrition, micronutrient deficiencies, overweight, and obesity. According to FAO (2013), the social and economic costs of malnutrition are enormous, amounting to around 3.5 trillion dollars per year. Of the 2.2 million global child deaths in 2004 that are attributable to underweight, almost half occurred in the African region, and more than eight hundred thousand occurred in the Southeast Asia region (WHO, 2009, p. 13). In other words, maternal and child malnutrition is more expansive than overweight and obesity, although the latter is increasing even in developing countries. Consequently, today, "65% of the world's population live in a country where overweight and obesity kills more people than underweight (this includes all high-income and most middle-income countries)" (WHO, 2009, p. 16). However, there are important local differences, because people living in the Global North have access to better public health systems.

Such food-related problems do not end at the consumption phase of these products. About one third of the food that is produced for human consumption is wasted. Specifically, the FAO has estimated that 95–115 kg/person is wasted yearly by European and North American consumers. The figures for sub-Saharan Africa and South and Southeast Asia are 6–11 kg/person yearly (ISWA, 2013, pp. 3–4). This volume of food waste contributes 4–11 percent of the world's greenhouse gas emissions (ISWA, 2013, p. 2). Moreover, food losses occur throughout the food chain. At the first stage, food is lost at the farm and during postharvest, mainly due to weather conditions, pest infestations, or the food's failure to meet quality standards. At the second stage, processing and wholesaling, the common causes are improper storage, improper transportation, and improper handling. At the retail stage, overstocking and damaged packing are among the main causes of loss. Finally, at the level of consumers and food services, the losses have several causes, including overpreparation. However, there are remarkable differences between developed and underdeveloped countries when it comes to wasted food.

In Europe, about 30 percent of a consumer's food produces waste. In North America, that waste estimate is closer to 40 percent. In industrialized Asia, the figure is 28.5 percent, and in developing countries it is around 15 percent (ISWA, 2013:5). A big problem is that for every kilogram of food produced, 4.5 kg of carbon dioxide is released into the atmosphere (ISWA, 2013, p. 5). Moreover, food waste produces methane, a gas that is even worse for the environment than carbon dioxide (FAO, 2011). In poor countries, the main causes of food waste are financial, technical, and managerial limitations of smallholder farmers. In developed countries, the consumer seems to be the main cause of waste. And on a per-capita basis, much more food is wasted in developed countries than in developing ones. Considering that millions of people are suffering from hunger in the world, this is an unreasonable and even criminal situation; the rich part of the world is wasting food that the poor part of the world urgently needs.

Conclusions and Observations

This chapter demonstrates the main features and consequences of the industrialization and globalization of agriculture, focusing particularly on the strong links in the realm of food between globalization and place-based dynamics. As we have seen, globalization and local conditions are mutually dependent forces. Moreover, there are important social movements and political initiatives promoting localism. The current climate is very dynamic, and, probably, the local will take new and varied shapes in the near future; but the outcome is unknown. On the other hand, neoliberal globalization has more than one single future. First, it has produced serious environmental, financial, and social problems, which together have produced a large number of victims and discontents. Second, a different kind of globalization is being promoted by a wide array of social and political forces in the North as well as in the South. At the same time it must be remembered that localism also has two faces: a positive one, discussed here, but also a negative aspect, pointed out by Goodman, DuPuis, and Goodman (2012). The local can be a space for freedom and justice, but it can also become a space of injustice, racism, privileges, discrimination, and monopolization. This means that the local is not positive or desirable per se, but it must be democratically constructed.

As we have seen in this chapter, industrialized agriculture makes enormous use of oil and its derivatives, and it cannot function without agrochemicals. Transporting and marketing food products require massive amounts of fossil fuels. The excessively high levels of carbon dioxide expelled by different sorts of equipment employed at every stage of the food-value chain illuminate the extent of these problems. Scarcity of food items is not the cause of the food crisis. On the contrary, actual food production is enough to satisfy twice the current food demand of the world population (How much is enough, 2011). The real causes are deeply rooted in the monopolistic and oligopolistic market structures at different levels of the global value chain—structures that make it impossible for large populations to have access to healthy food products at affordable prices. Also relevant is the fact that large populations are often excluded from the positive aspects of economic and social modernization. In other words, the real food problem is not a productivity issue but a matter of fair distribution of wealth, resources, capabilities, opportunities, and rewards.

It is important to emphasize that the big three regime crises—food, peak oil, and climate—are inherently linked, as Shiva (2013) convincingly argues, and they must be solved together. Their effects are global, but their causes are forged under local circumstances. Accordingly, their solutions must be constructed from the bottom, at local levels, but to be fully effective, efforts must be generalizable and congruent at the global level as well. Accordingly, it is clear that many issues must be addressed at different levels, including industry, urbanization, transport, education, health services, democracy, gender equality, etcetera. However, the first steps must be taken in the food realm, because if the present ways of producing, eating, and conceptualizing food are not modified radically, nothing will really change. The industrialized production of food is unsustainable and harmful for human and nonhuman beings, and even for the planet. The "modern" diet—plenty of meats, sugar, salt, and fat—is unhealthy and very expensive from a social point of view, considering the costs for the treatment of diseases caused by this diet. Besides, excessive buying of food items by wealthy populations contributes to increasing food waste.

Finally, and perhaps most importantly, we must change our "modern" idea of food, which might be accomplished in two ways. First, food is not just merchandise to be freely bought and sold on an open market. Rather, it is an essential condition of life, and therefore access to it should be a right enjoyed by everyone in every part of the world. Second, food is not just a thing but a privileged way of communication between human beings dwelling in different places and between human and nonhuman beings around the earth. Moreover, culturally speaking, food is a way of communication between past, present, and future generations, whose destiny is profoundly connected to local culture, structure, and place-based dynamics.

References

Arce, A., & Marsden, T. K. (1993). The social construction of international food: A new research agenda. *Economic Geography, 69*(3), 293–311.

Bagcchi, S. (2014). Hypothyroidism in India: More to be done. *Lancet. Diabetes & Endocrinology, 2*(10), 778.

Bailey, R. (2011). Growing a better future: Food justice in a resource-constrained world. *Oxfam.* Retrieved from http://www.oxfam.org/sites/www.oxfam.org/files/cr-growing-better-future-170611-en.pdf

Belasco, W., & Horowitz, R. (Eds). (2009). *Food chains: From farmyard to shopping cart.* Philadelphia: University of Pennsylvania Press.

Beraud, J. L., Galindo, J. G., & Covantes, C. (Eds.). 2008. *Jornaleros y medio ambiente: Los agroquímicos en la agricultura sinaloense.* Culiacán, Mexico: Universidad Autónoma de Sinaloa.

Brown, L. (2012). *Full planet, empty plates: The new geopolitics of food scarcity.* New York, NY: W. W. Norton & Company.

Buci-Glucksmann, C. (1980). *Gramsci and the State.* London: Lawrence and Wishart.

Burch, D., & Lawrence, G. (2009). Towards a third food regime: Behind the transformation. *Agriculture and Human Values, 26*(4), 267–279.

Canavari, M., & Olson, K. D. (Eds.). (2007). *Organic food: Consumers' choices and farmers' opportunities.* New York, NY: Springer.

Carías, P. (2014, January 1). ¿Qué está matando a los agricultores? El Faro. Retrieved from http://www.elfaro.net/es/201401/noticias/14366/

Comisión Económica para América Latina y el Caribe, Organización de las Naciones Unidas para la Agricultura y la Alimentación, y del Instituto Interamericano de Cooperación para la Agricultura (CEPAL/FAO/IICA) (2011). *Boletín Perspectivas de la agricultura y el desarrollo rural en las Américas: Una mirada hacia América Latina y el Caribe.* No. 1. Volatilidad de precios en los mercados agrícolas (2000–2010): Implicaciones para América Latina y opciones de políticas. Comosión Económica para América Latina: Santiago de Chile.

Dolan, C., & Humphrey, J. (2004). Changing governance patterns in the trade in fresh vegetables between Africa and the United Kingdom. *Environment and Planning, A36*(3), 491–509.

Eddleston, M., Karalliedde, L., Buckley, N., Fernando, R., Hutchinson, G., Isbister, G.,... Smit, L. (2002). Pesticide poisoning in the developing world—a minimum pesticides list. *Lancet, 360*(9340), 1163–1167.

ETC Group (2008). *Who qwns nature? Corporate power and the final frontier in the commodification of life.* Communiqué number 100. Retrieved from http://www.etcgroup.org/content/who-owns-nature

European Parliament (2007). Policy department structural and cohesion policies. *The gap between producer prices and the prices paid by the consumer.* Agriculture. Study IP/B/AGRI/IC/2007_001 PE 397.240 E October.

Fairtrade Canada (2012). Socially conscious consumer trends fair trade: Market analysis report / April. Retrieved from http://www5.agr.gc.ca/resources/prod/Internet-Internet/MISB-DGSIM/ATS-SEA/PDF/6153-eng.pdf

Food and Agriculture Organization (FAO). (2011). Global food losses and food waste: Extent, causes and prevention. Retrieved from http://www.fao.org/docrep/014/mb060e/mb060e.pdf

Food and Agriculture Organization (FAO). (2013). *FAO statistical yearbook 2013: World food and agriculture.* Retrieved from http://www.fao.org/docrep/018/i3107e/i3107e.pdf

Ferran, F., & Grunert, K. G. (2007). French fair trade coffee buyers' purchasing motives: An exploratory study using means-end chains analysis. *Food Quality and Preference, 18*(2), 218–229.

Francks, P. (2009). *The Japanese consumer: An alternative economic history of modern Japan*. Cambridge: Cambridge University Press.

Friedland, W. H. (2001). Reprise on commodity systems methodology. *International Journal of Sociology of Agriculture and Food, 9*(1), 82–103.

Friedmann, H. (1982). The political economy of food: The rise and fall of the postwar international food order. *American Journal of Sociology, 88*, S248–S286.

Friedmann, H. (1993). The political economy of food: A global crisis. *New Left Review, 1*(197), 29–57.

Friedmann, H. (2009). Discussion: Moving food regimes forward: Reflections on symposium essays. *Agriculture and Human Values, 26*(4), 335–344.

Friedmann, H., & McMichael, P. (1989). Agriculture and the state system: The rise and decline of national agricultures, 1870 to the present. *Sociologia Ruralis, 29*(2), 93–117.

Gereffi, G., & Korzeniewicz, M. (Eds.). (1994). *Commodity chains and global capitalism*. Westport, CT: Praeger.

Gereffi, G., & Christian, M. (2009). The impacts of Wal-Mart: The rise and consequences of the world's dominant retailer. *Annual Review of Sociology, 35*(1), 573–591.

González, L. (1974). *San José de Gracia: Mexican village in transition*. Austin: University of Texas Press.

Goodman, D., DuPuis, E. M., & Goodman, M. K. (2012). *Alternative food networks. Knowledge, practice, and politics*. London: Routledge.

Goodman, D., & Watts, M. (Eds). (2013). *Globalising food: Questions and global restructuring*. London: Routledge.

Gottdiener, M. (Ed.) (2000). New forms of consumption: Consumers, culture, and commodification. Lanham, MD: Rowman & Littlefield.

Gutmann, M. C. (2000). Por quien doblan las campanas de Taco Bell: Respuestas populares al TLC al sur de la frontera. *Alteridades, 10*(19), 109–122.

Higman, B. W. (2012). *How food made history*. West Sussex: Wiley-Blackwell.

Hobsbawm, E. (1994). *Age of extremes. The short twentieth century 1914–1991*. London: Abacus.

How much is enough? (2011). *Economist, 399*(8733), 10–13.

Huang, S. (2004). An overview of global trade patterns in fruits and vegetables. In S. Huang, L. Calvin, W. T. Coyle, J. Dyck, K. Ito, D. Kelch,... & T. Worth, *Global trade patterns in fruits and vegetables* (pp. 3–15). USDA Economic Research Service. Retrieved from http://www.ers.usda.gov/media/320468/wrs0406c_1_.pdf

Humphrey, J. (2007). The supermarket revolution in developing countries: Tidal wave or tough competitive struggle? *Journal of Economic Geography, 7*(4), 433–450.

International Solid Waste Association (ISWA). (2013). *Food waste as a global issue—From the perspective of municipal solid waste management.*

Retrieved from http://waste.ccac-knowledge.net/documents/iswa-key-issue-paper-food-waste-global-issue-perspective-municipal-solid-waste-management

Kritzinger, A., Barrientos, S., & Rossouw, H. (2004). Global production and flexible employment in South African horticulture: Experiences of contract workers in fruit exports. *Sociologia Ruralis, 44*(1), 17–39.

Lee, H., Summer, D., & Vergati, J. (2012). Farm price margins constructed under alternative price calculation. *California Department of Food and Agriculture*. Retrieved from http://aic.ucdavis.edu/MarketingMargins/documents/ReportOnFarmMargins.docx

Magdoff, F., & Tokar, B. (Eds.). (2010). *Agriculture and food in crisis: Conflict, resistance, and renewal*. New York, NY: Monthly Review Press.

Magdoff, F., Foster, J. B., & Buttel, F. H. (Eds.). (2000). *Hungry for profit: The agribusiness threat to farmers, food, and the environment*. New York, NY: Monthly Review Press.

McMichael, P. (2009a). A food regime analysis of the "world food crisis." *Agriculture and Human Values, 26*(4), 281–295.

McMichael, P. (2009b). A food regime genealogy. *Journal of Peasant Studies, 36*(1), 139–169.

Meyer, E., & Bernier, I. (2010). *Standardizing or adapting the marketing mix across culture. A case study: Agatha*. Thesis in marketing. Halmstad University, School of Business and Engineering.

Moreno, J., & López, M. (2005). Desarrollo agrícola y uso de agroquímicos en el Valle de Mexicali. *Estudios Fronterizos, 6*(12), 119–153.

Morgan, K., Marsden, T., & Murdoch, J. (2006). *Worlds of food: Place, power, and provenance in the food chain*. Oxford: Oxford University Press.

Ongley, E. (1996). *Control of water pollution from agriculture*. FAO irrigation and drainage paper 55. Rome: Food and Agriculture Organization of the United Nations.

Organisation for Economic Co-operation and Development (OECD). (2007). *Agricultural policies in OECD countries. Monitoring and evaluation 2007*. OECD: Paris.

Pilcher, J. M. (2006). *Food in world history*. New York, NY: Routledge.

Polanyi, K. (1944). *The great transformation: The political and economic origins of our time*. New York, NY: Rinehart.

Pretolani, R., Cavicchioli, D., & Cairo, V. (2014). Marketing margins of food products in European countries using input-output tables. Working paper N. 2014–06, April. Milan: Universitá degli Studi di Milano.

Riojas-Rodríguez, H., Schilmann, A., López-Carillo, L., & Finkelman, J. (2013). Environmental health in Mexico: Current situation and future prospects. *Salud Pública México, 55*(6), 638–649.

Sale, K. (1993). *The green revolution: The American environmental movement, 1962–1992*. New York, NY: Hill & Wang.

Schlanger, Z. (2014, July 24). Exposure to pesticides when pregnant linked to 3 generations of disease. *Newsweek*. Retrieved from http://

www.newsweek.com/pesticide-diseases-pesticide-exposure-pesticide-poisoning-pesticides-and-pregnancy-261181

Schnept, R. (2013). Farm-to-food price dynamics, Congressional Research Service 7–5700, CRS Report for Congress R40621. Prepared for Members and Committees of Congress. Retrieved from http://fas.org/sgp/crs/misc/R40621.pdf

Segrelles, J. A. (2001). Problemas ambientales, agricultura y globalización en América Latina. *Scripta Nova*. Universidad de Barcelona. No. 92, July. Electronic journal. Retrieved from http://www.ub.edu/geocrit/sn-92.htm

Shirai, M. (2010). Analyzing price premiums for foods in Japan: Measuring consumers' willingness to pay for quality-related attributes. *Journal of Food Products Marketing*, *16*(2), 184–198.

Shiva, V. (1992). *The violence of green revolution: Third world agriculture, ecology and politics.* London: Zed Books.

Shiva, V. (2000). *Stolen harvest: The hijacking of the global food supply.* Cambridge, MA: South End Press.

Shiva, V. (Ed.). (2007). *Manifestos on the future of food and seed.* Cambridge, MA: South End Press.

Shiva, V. (2013). *Soil not oil: Environmental justice in an age of climate crisis.* London: Zed Books.

Smil, V., & Kobayashi, K. (2012). *Japan's dietary transition and its impacts.* Cambridge, MA: MIT Press.

Smith, C., Kerr, K., & Sadripour, A. (2008). Pesticide exports from U.S. ports, 2001–2003. *International Journal of Occupational Environmental Health 14*(3), 176–186.

Staertzel, L. (n.d.). Carrefour in China: A crossroad between East and West. Retrieved from http://bear.warrington.ufl.edu/oh/IRET/Student%20Experience/Carrefour%20Essay%20Edited.pdf

Standage, T. (2009). *An edible history of humanity.* New York, NY: Walker Books.

Stevenson, M. (2007, October 10). Taco Bell's fare baffles Mexicans. *Seattle Times.* Retrieved from http://seattletimes.com/html/businesstechnology/2003937804_tacobell100.html

Valdez, S. B., García, E., & Wiener, M. S. (2000). Impact of pesticides use on human health in Mexico: A review. *Reviews on Environmental Health, 15*(4), 399–412.

Watson, J. L., & Caldwell, M. L. (Eds.). (2005). *The cultural politics of food and eating: A reader.* Malden, MA: Blackwell.

Wilk, R. (Ed.). (2006). *Fast food/slow food: The cultural economy of the global food system.* Lanham, MD: AltaMira Press.

Willer, H., Yussefi, M., & Sorensen, N. (Eds.). (2008). The world of organic agriculture: Statistics and emerging trends 2008. Frick: International Federation of Organic Agriculture Movements (IFOAM), and Research Institute of Organic Agriculture (FiBL). Retrieved from http://orgprints.org/13123/4/world-of-organic-agriculture-2008.pdf

World Health Organization (WHO). (2014a). *Qualitative risk assessment of the effects of climate change on selected causes of death, 2030s and 2050s.* Simon Hales, Sari Kovats, Simon Lloyd, Diarmid Campbell-Lendrum, (Eds.). Geneva: World Health Organization of the United Nations:

World Health Organization (WHO). (2014b). *World Health Statistics.* Geneva: World Health Organization of the United Nations.

World Trade Organization (WTO). (2012). International Trade Statistics. Retrieved from http://www.wto.org/english/res_e/statis_e/its2012_e/its12_merch_trade_product_e.htm

Yaqub, F. (2014, May 24). Kidney disease in farming communities remains a mystery. *Lancet*, *383*(9931), 1794–1795.

Chapter 3

Lessons from the Past: A Historical Overview of the Place-Food Relationship

Mark B. Tauger

Introduction

Food may be transported, but agriculture depends upon the local environment. Consequently, many, if not most, disasters in the histories of agrarian systems have also been the result of local environmental events or of decisions by farmers and political leaders to ignore local conditions and risks. This chapter examines a few indicative cases of such disasters as historical examples of the importance of the relationship between food and place. It also sheds light on the fundamental shift in this relationship, which occurred in the modern period, and its implications for the future.

Practitioners dealing with food crises might question the value of reviewing past disasters, assuming that modern economic and scientific knowledge has left all those mistakes behind. Yet historians argue, as the historian Peter Stearns wrote for the American Historical Association, that the study of history provides "access to the laboratory of human experience" and broadens our awareness of events (Stearns, 1998). The economist Steven Kates argues for the value of studying economic history and the history of economic thought because unlike the natural sciences, economics is a social science and, as such, is subject to revision and change in light of new perspectives both on current and on past events (Kates, 2013, pp. 1–20). The history of agrarian crises and their relations to particular localities provide considerable evidence that can be applied to recent and current events, especially because some of the most important cases are quite recent and the causes of them are still operating.

There are two main and overlapping historical periods in the history of disasters involving food and place. In the first period, from antiquity to approximately 1300, most societies were self-sufficient and self-contained in agriculture. They traded food and other agricultural products, but trade played a secondary part in the subsistence of the great majority of these populations. In the second period, from the late Middle Ages until the present, increasing trade led to the production of agricultural products for export, especially after the Black Death and the European conquest and assimilation of the Americas and parts of Asia. High-consuming countries became increasingly dependent on imported food. During the twentieth century, many countries in Asia, Africa, and Latin America came under foreign agricultural influences through colonialism and development policies.

This shift from agricultural autarky to agricultural interdependence also changed the role of place in food and agricultural crises. In the earlier period, agricultural crises resulted mainly from large-scale environmental disasters and long-term agrarian practices. This history shows the importance of anticipating such environmental events and understanding the long-term impacts of farming practices and other activities that affect farming.

In the more recent period, while large-scale and long-term practices continued to cause crises, short-term human actions—especially government policies and private-sector actions—played an increasingly important role in disasters. Yet this era also saw significant scientific advances that made it possible to anticipate crises much better. As a result, disasters often resulted at least in part from political convenience, corporate greed, and other short-term causes. This record in turn implies the importance of ensuring that governments and businesses are aware of the risks inherent in their activities and are accountable for their policies.

Ancient and Medieval Disasters

The earliest cases of major disasters are uncertain and are based on extrapolations from archaeological data, but more recent cases are much better documented. Three examples of early agrarian crises—the collapse of the Mayan civilization of Central America, the flooding of the north China rivers, and the famine and plague in fourteenth-century medieval Western Europe—were similar in their large-scale environmental causes but different in the economic, social, and political responses of the affected societies.

The Mayan Collapse

The Maya lived in a region stretching from the Yucatan Peninsula to southern Guatemala. The region was lush but had a monsoon pattern of wet and dry seasons (Diamond, 2005). While the sources are mostly archaeological, it is clear that the Maya grew the whole range of classic Central American food crops, including corn, beans, squash, chilies, sweet potatoes, manioc, avocados, tomatoes, and other plant foods. They also had deer, turkey, iguana, and other small animals, fish, and insects (Coe, 1994). From about 600 BCE buildings and human remains evidence a growing civilization. By 250 BCE they had villages and towns all over the Guatemalan lowlands and at least five major towns with large temples and irrigation canals. Despite some wars and destruction, by 250 CE the Maya had a large civilization with major cities, buildings, roads, writing, complex sculptural works, and an elaborate social structure with tens of thousands of commoners supporting a warlike aristocracy that lived in luxurious palaces. The Maya cut down vast areas of forest to support farming and the mining of lime and stone for construction (Adams, 1997).

From about 600 CE a series of wars and population growth imposed demands for food that exceeded the crop and animal production and sanitation capabilities of the society. Archaeological remains from 600–900 CE show evidence of widespread malnutrition and disease. Chronic warfare, possibly over resources and food, further exhausted the civilization. Then, during the eighth through the tenth centuries, a series of prolonged and severe droughts devastated Mayan food production. The famines and conflicts associated with them killed a large share of the population (Adams, 1997). The survivors returned to subsistence farming and did not attempt to reconstruct the civilization, and the region returned to rain forest until settlement revived in the twentieth century (Iannone, 2014).

The Mayan collapse is widely viewed as a classic example of the consequences that follow when a society ignores its environmental limits, fails to anticipate and prepare for an environmental disaster, and devotes too much of its resources to war (Diamond, 2005). While the Maya had little modern scientific knowledge, the archaeological evidence shows that the ninth-century drought was not the first in their history and that they had recovered from earlier ones. This evidence implies that at least in part, the Maya collapse resulted from their failure to remember and understand their own past, perhaps under the pressure of internecine war in the centuries preceding the great drought.

China's Response to Environmental Crises

China has had numerous agrarian crises in its history, but in China's early history most of its disasters resulted from floods as much as from drought. The early Chinese civilization emerged in the northern plains, which were supplied with water by the Yellow River and other rivers. The Yellow River earned that name from the silt that flowed into it from the loess soils of northern China. During some four thousand years of China's history up to 1900, Chinese farmers cut down large areas of forest in the loess region and expanded their croplands. These practices increased the siltation of the region's rivers. The increased siltation raised the riverbeds and resulted in frequent flooding (Elvin, 2004). Chinese rulers imposed taxation in labor to build levees and diversions to restrain floods, but periodically, extreme snow and rainfall at the rivers' sources in Mongolia and western China caused enormous water flows that overwhelmed the protective measures, causing over a thousand floods that in dozens of cases changed the course of the Yellow River (Chen, Syvitski, Gao, Overeem, & Kettner, 2012). In several cases the floodwaters did not recede over the region for many years, causing millions of deaths.

Unlike the Maya, however, the Chinese had a civilization that in theory—and very often in practice—placed a high value on farmers as the economic and social basis of their society (Deng, 1999). Also, for most of its history, China was a unified state and did not have as many internecine conflicts as the Maya. Chinese governments often stockpiled grain to feed and provide seed for peasants displaced by disasters, deployed numerous trained and experienced officials to report on these crises to the central government and to manage relief efforts, and often organized large-scale reconstruction efforts to overcome natural disasters and restore farming (Will & Wong, 1991). China's responses to these crises, in times when dynasties were at their peak, were the most effective responses to crises in any country in the world in medieval times. These crises both reflected and contributed to the strength of the Chinese state. Despite the dislocations they caused, the vast floods restored the fertility of the lands, and once the waters retreated, peasants returned and farmed in essentially the same ways as before.

The Chinese peasants' relationship to their lands in the north Chinese farming system thus resembled the Mayan peasants' relationship to place. As population grew in the Ming and Qing dynasties (1368–1911), farmers cleared yet more forests and brush, despite appeals by the government to stop, and laid the groundwork for more floods, famines, and crises (Li, 2007). The Chinese differed from the Maya, however, in certain key ways. First, their society's respect for farmers and its relief and recovery programs enabled the northern regions to recover at least partially from crises. Second, China, early on, had specialists in the government and outside it, who studied and wrote books and pamphlets on agriculture to inform readers about how to improve soils and reduce the risks of disasters (Bray, 1984). These gradually brought about significant improvements in Chinese farming, despite the disasters. Finally, many Chinese moved south to the Yangtze River valley and delta, often with government support. In that region, floods were less frequent and severe than in the north.

Europe's Fourteenth-Century Crisis

A third example of an ancient-medieval agrarian crisis that resulted from large-scale environmental disasters was the combination of the great famine of 1315–1317—the effects of which lasted into 1322 (Kershaw, 1973)—and the Black Death of 1347–1349 in Western Europe. These disasters followed three centuries of economic and agricultural expansion and population growth that pushed subsistence farming into marginal and submarginal lands and created a growing population of impoverished people in towns and villages. This expansion took place during the medieval optimum period of relatively warm climate. But by 1300, Western Europe overall was approaching a Malthusian crisis in which food production barely kept up with the needs of a growing population and crop failures began to cause famines (Duby, 1968).

The crisis began with wet weather in 1314, which reduced grain harvests and drove up prices (Jordan, 1996). Then in May 1315, extremely heavy rains began and continued without interruption all summer and into the fall across the British Isles, northern Europe from Spain to Scandinavia and Russia, and much of southern Europe as well. The flooding, which recurred in 1316 and was only partly alleviated by dryer conditions in 1317, reduced and sometimes widely destroyed crops, stimulating the spread of ergotism (a fungal plant disease that can kill people) and other epidemics. The rain and flooding also created extremely favorable conditions for the vast spread of livestock disease, an epizootic that killed vast numbers of sheep and nearly two-thirds of the cattle in England and Wales and certain other parts of Europe between 1319 and 1320. Rains and floods alternated with droughts, and famines and livestock deaths recurred in subsequent years. Peasants, serfs, and other poor people, already short of food in 1300, suffered a mortality rate of 10 percent or more between 1315 and 1317. Food prices doubled and many monasteries and landlord estates became impoverished, but some big landlords profited from the high prices even though their harvests were much less than normal. The King of England demanded taxes from the peasants and the poor, but he gave some money to impoverished monasteries.

This great famine was only the beginning of an even larger crisis (Aberth, 2010). Floods and droughts recurred, and famines and epizootics with them, during the following decades. These had the long-term effect of physiologically weakening the population, making it more vulnerable to the great Black Death epidemic that spread from Asia to northern Italy in 1347 and to the rest of Europe during the following two years. While this was not an agrarian event, it had dramatic effects on agriculture. Because it killed between a fourth and a third of Europe's population, the vast majority of whom were peasants, serfs, or those otherwise involved in the production of food and fiber, the plague immediately reduced the number of farmers and farm laborers, and thereby food production for many years.

The crisis of the early fourteenth century had complex long-term results, including the weakening of Western European serfdom and the rise of serfdom in Eastern Europe. But the biggest change from an agrarian viewpoint was a significant shift in European food production that became evident a few decades after the Black Death, when Europe began to recover. Whereas many subsistence farmers remained, farmers in the Netherlands, northern France, and other regions—where cities revived increasingly—shifted from growing grain crops to raising livestock and more valuable vegetable crops for expanding urban markets (Bautier, 1971; Duby, 1968). These regions began to obtain their grain supplies from Eastern Europe, from the newly emerging serf estates in the Baltic region. This new demand led to the "second serfdom" in Eastern Europe and to the Atlantic system of slave plantations, both of which emerged in the sixteenth century (Blum, 1957).

These three examples show different ways in which early societies responded to agricultural and food crises in particular places. The Maya apparently exhausted their agricultural environment and lacked the knowledge to anticipate or ameliorate the results and consequently disappeared as a civilization. Chinese officials and farmers approached crises in a rational, pragmatic and humanitarian manner, recorded and passed on their experience in dealing with natural disasters, and by the late-medieval period, had expanded their civilization to the south and begun a major period of growth and innovation. China also had far greater resources than did the Maya because of the large size of the Chinese state and its overall agricultural productivity. Europe in the fourteenth century also had considerable resources, but the combination of climate disasters, crop failures and famines, and epidemics caused serious devastation that took the region almost a century to overcome. While the European governments did not respond with anything like the humanitarian and skilled responses of the Chinese, European farmers adapted to new market conditions.

Modern Crises of Food and Place

In the modern period, several agrarian crises affected almost the whole northern hemisphere. These included the Little Ice Age that brought extreme cold, crop failures, and famines, and contributed to epidemics chronically from the fifteenth to the eighteenth centuries, and some singular disasters. From 1601 to 1603, dust from the eruption of the Huaynaputina volcano in 1600 blocked sunlight widely over the northern hemisphere, causing extremely cold winters and cool rainy summers for all these years, which in turn greatly reduced grain harvests in Russia (as well as elsewhere in the northern hemisphere) and resulted in one of the longest and most severe famines in Russian history. In 1709 and 1740, extreme cold and heavy snow and rain caused by the Little Ice Age climate fluctuations destroyed crops by winterkill and froze rivers, which hindered transport of famine relief in southern France and elsewhere across Europe (Monahan, 1993). The volcanic eruption of Tambora in Indonesia in 1815 caused similar problems. In these, as in earlier events, people did not have the scientific knowledge to understand the underlying causes of the disasters, but governments and local groups did attempt to provide some famine relief (Fagan, 2002).

In other cases, people's actions led to agricultural crises, and in many of these cases the actors eventually understood their own share of responsibility for these events. The most important agricultural transformation in the modern period, and the one that brought many of the largest disasters, was the agricultural assimilation of the global grasslands, which took place in the nineteenth and twentieth centuries. The most significant aspects of this vast effort included the Russian conquest of the black-earth steppes, the settlement of which began in the 1830s, the almost-simultaneous British colonial movement into the Australian grasslands, the American and Canadian farmers' migration into the Great Plains and west coast after the Civil War, and the European assimilation of the Pampas of Argentina that began in the 1880s. Similar patterns occurred on smaller scales elsewhere, as in South Africa (Adelman, 1994; Meinig, 1962; Moon, 2013; Worster, 1979).

All these regions usually had arid climates, but their weather often alternated between periods with reasonable rainfall and those of severe drought or other harsh weather conditions. These changing patterns could mislead settlers. For example, American farmers who came to the plains in the 1870s and the 1880s encountered relatively mild arid weather that did not prepare them for the extreme droughts of the 1890s. The problem of weather soon became evident to farmers and to the newly emerging groups of educated agricultural scientists, but it took many years, long-term research, and several crises before the different groups involved learned to take local ecology into account.

Agricultural Crisis in the Soviet Union

Russian peasants and landlords were the first to farm the great Eurasian steppes, in what is now south Russia, Ukraine, the Volga basin, the southern Urals, and southwest Siberia and northern Kazakhstan. Early on they had to deal with drought, flooding, wind, and erosion. Sometimes merely plowing the soil in some of these regions dried it out and caused dust storms and crop failures. The Russian government ultimately spent millions of rubles on famine relief and other aid to peasants. A few Russians, at first landlords with farming experience and then scientifically educated specialists and government officials, published analyses, studies, and advice about how to deal with the chronic weather crises. Some of this research, such as Vasilii Dokuchaev's work on soils, made substantial contributions to world science. A few groups of farmers developed effective methods of resisting drought. German Mennonite settlers in the steppe regions used "black fallow," the repeated plowing of fields to prevent grass and weed growth, to retain moisture and to obtain higher grain yields from those fields (Moon 2013; Tauger, 2011b).

Russian scientists and other specialists publicized their views about steppe farming, and some peasants in the late nineteenth and early twentieth centuries began to apply improved farming methods such as black fallow with generally good results. But most of them continued to use traditional extensive-farming methods that basically wagered that the weather would be favorable rather than employing methods that would conserve soil and accommodate potential harsh weather. Russia could produce large grain harvests and compete with the United States in exports, but it also had many crop failures and famines that required government relief.

This pattern of extensive farming and weather-induced crop failures continued into the Soviet period; civil war and communist policies intensified the effects of natural disasters. During the period 1920 to 1921 and the years 1924 and 1928—the period of the "New Economic Policy" or NEP, when the USSR attempted to recover from World War I and the Russian Civil War—the Volga region, Ukraine, and other regions endured a series of extreme weather events, including droughts, winterkill, plant diseases, and insect infestations. These resulted in famines, which reached their peak during the period 1921 to 1923, causing millions of deaths beyond normal levels. In response, the Soviet government repeatedly imported food, established famine-relief agencies, and distributed food, seed, forage, and equipment. From 1922 to 1923, the Soviet government even allowed the American Relief Administration to provide aid. In 1924, the regime mounted another relief program, including food imports, in response to drought, and during the period 1925 to 1926, the previous years' relief supported a significant recovery. The recurrence of extreme weather in 1928, followed by another relief campaign, persuaded Soviet leaders that the backwardness of peasant farming was the cause of these chronic crises. On the basis of this viewpoint, the regime undertook the vast campaigns of collectivization and dekulakization (the exile outside the village of oppositional "better off" peasants) to transform Soviet peasant farming into what they hoped would be a Soviet communist version of American mechanized agriculture (Tauger, 2001; Tauger, 2006; Tauger, 2011a).

The 1928 crop failure presaged a series of bad years, culminating in two extremely poor harvests in 1931 and 1932, which led to one of the worst famines in Russian history. The regime provided relief from its own reserves, but for political reasons Soviet leader Joseph Stalin refused to import food, unlike in previous years. More than four million people died from this famine before the large 1933 harvest ended it. A similar crisis occurred from 1946 to 1947, before the country had recovered from the devastation of World War II, and again the regime—for political reasons—refused to import food. This time about two million people died of famine before the following year's harvest could save them (Tauger, 2011b).

This story of great disasters, however, is incomplete: the other part of the story concerns the efforts by Soviet agricultural scientists, economists, and other specialists to find solutions to these problems.

Simultaneously with collectivization, the Soviet regime established a network of agricultural research stations headed by a central agricultural research center under the Soviet scientist Nikolai Vavilov. Most studies of this aspect of Soviet history reduce it to the damaging effects of the pseudoscientist Trofim Lysenko.

Vavilov was a plant geneticist who specialized in plant immunity to disease, and he was the head of the Soviet Academy of Agricultural Sciences from its foundation in 1929 until he was ousted by Lysenko in 1936. He followed conventional genetics and enjoyed enormous international respect. Lysenko was a poorly educated agronomist in Ukraine, who took the wrong side in the 1920s debate over inheritance, and rigidly subscribed to the idea of the inheritance of acquired characteristics. But Lysenko made all kinds of exaggerated promises that persuaded Stalin to support him over Vavilov, which led to Vavilov's arrest in 1940 and death in prison and to Lysenko's rise to the top post in agricultural research in 1948.

The damage that Lysenko caused was actually relatively limited, however, because many scientists managed to evade his attempts to control research. Vavilov had educated a generation of scientists who kept his memory and ideas alive after his death despite Stalin and Lysenko (Pringle, 2008; Roll-Hansen, 2005). Many of these scientists continued to do legitimate research and tried to help Soviet farmers improve production. The key figure was the plant breeder Pavel Luk'ianenko from the Kuban region east of the Black Sea. Inspired in part by Vavilov, Luk'ianenko developed the first Soviet semi-dwarf high-yielding varieties of wheat in the 1950s, beginning a Soviet Green Revolution that substantially raised Soviet farms' grain yields in years of good weather (Tauger, 2012). Soviet scientists such as Luk'ianenko continued the work of earlier improving landlords and Russian scientists. But Russian and Soviet scientists disagreed over how to help farmers manage the droughts, infestations, and other problems that they faced in the Eurasian steppe.

Meanwhile, Soviet leaders sought to increase food production at all costs to have some reserves in case of crop failure and famine. The approach to farming promoted by collectivization was basically a restructuring of land use and the introduction of small tractors of that era. But after Stalin's death, with the USSR facing another major shortfall in food supplies, the new Soviet leader, Nikita Khrushchev, proposed the "Virgin Lands" program to grow grain on some twenty-five million hectares in western Siberia and Kazakhstan, the southern Urals, and the east side of the Volga. All these areas were arid lands. The leaders basically saw this as a stopgap to overcome a short-term shortage. The Stalin regime had actually tried this project between 1928 and 1930, with limited success. In 1954, Khrushchev revived the program and the Virgin Lands became a core and productive part of Soviet agriculture. But from 1962 to 1963, after several years of adequate-to-good harvests, the rains failed again, and wind caused dust storms that blew away millions of tons of topsoil from lands plowed extensively in the previous years. Khrushchev had to import ten million tons of grain from Canada and the United States to cover the shortfall normally covered by the Virgin Lands (McCauley, 1976).

This debacle could be interpreted as an example of people ignoring environmental conditions, but it was more complex than that. Both in 1928 and in 1954, experts debated the proposals to farm in these regions, and they differed. While some emphasized the risk of drought, others thought it would be possible to use the region in the short term. In historical perspective, the Virgin Lands campaign resembles traditional Russian peasant farming—a gamble taken in planting a large area, in the hope that at least part of the area would produce a crop (analogous perhaps to a modern mutual fund). In the wake of the crop failures of 1972 to 1975, Soviet scientists conducted more research, studied other countries' research, and proposed methods to make it possible to continue farming these regions; and they were reasonably successful despite periodic dry years and crop failures, and ill-advised large-scale projects imposed by the Brezhnev-era Soviet regime (Shabykina, 1976).

Independent Kazakhstan is still a major wheat grower and exporter, although Kazakh farmers cultivate less than half of the previously farmed Virgin Lands (eleven million out of twenty-three million hectares) because so much of the land was eroded and vulnerable (FAO, 2014). Droughts and disasters continued after the fall of the Soviet Union. For example, post-Soviet Russia expected to export large quantities of wheat in 2009, but severe drought caused wildfires in many grain fields, requiring President Vladimir Putin to cancel export plans in 2010 and again import food (Welton, 2011).

The Russian and Soviet series of disasters were derived from the extreme continental climate and other environmental hazards in the grasslands and the country's surrounding regions. "Place" in this case was a very large area, but it was relatively uniform in the problems that farmers faced there. The significance of place was that Russians and Soviets faced a chronically unstable environment: good years alternating with bad and very bad years. They responded to these circumstances with a similar alternation of gambles and pragmatic-scientific policies that brought some improvement, but chronic problems persist and Russia and Kazakhstan have had to abandon vast land areas because of soil depletion.

An American Agricultural Crisis of Place: The Dust Bowl

The American Great Plains and the Dust Bowl of the 1930s provide an example that resembles the Russian case in the environmental causative factors, with significantly different government and popular responses. American farmers began moving into the Plains in the 1850s, a migration that accelerated after the Civil War. Up into the 1880s, the cowboys and the stockbreeding companies who employed them dominated the region. After the great cattle die-off in the freezing winter of 1886, the grain farmers took over much of the Plains as well as much farmland along the west coast, from California to Washington. Farmers encountered drought repeatedly. The US Department of Agriculture (USDA) sent staffers to other countries, including Russia, to find drought-resistant crops. The agency also commissioned US agricultural specialists to study better farming methods, and they came up with "dry farming." The USDA, however, cautioned farmers—in several publications—that even these methods could not guarantee success (Hurt, 2002).

Still US farmers pushed their luck and sold crops for high prices during World War I. When grain and other food prices crashed in 1920, many farmers' financial circumstances became precarious and many failed outright. Farmers responded to these falling prices by producing more, and the arrival of the Fordson tractor in 1917 allowed them to plow and plant more land faster than ever before. They continued to cultivate more land despite the Great Southern Drought of 1930, the northern drought in the Dakotas and Montana in 1931, and dry weather in the Plains in subsequent years. But the intensive mechanized cultivation that farmers conducted for years on end left Plains soils dried out and eroded. When intense winds began to blow in May 1934, the Great Plains experienced dust storms, topsoil loss, and destruction of farms as never before. Dust storms damaged twenty-one million acres seriously and made a much larger area a bad farming risk. Overall, during the 1930s, farmers abandoned sixty-four million acres, while the US government set aside fourteen million acres for conservation as unsuited to farming. Many abandoned their devastated farms and took their families west, but some decided to stay and wait out the crisis. Farmers also organized themselves politically to persuade the government to introduce support policies, with little success until the administration of Franklin D. Roosevelt in 1933 (Worster, 1979).

Under Roosevelt, the USDA finally addressed the farm crisis with subsidy programs and research. In particular, the agency established a Soil Conservation Service to advise farmers, produced several publications suggesting contour plowing and other soil-conservation methods, and identified regions too eroded to farm, authorizing the government to pay farmers to remove land from production. All these measures were effective, but farmers continued to grow crops in the Plains, and drought between 1954 and 1957 produced dust storms that damaged sixteen million acres until rains returned. In 1977, dust storms again blew soil from Texas to Oklahoma, and yet again in 1988, when relief and soil protection cost the government $20 billion (Opie, 1992).

The discovery of the Ogallala aquifer that underlies 174,000 square miles of the Great Plains from Texas to South Dakota rescued farmers at least for the short term, starting in the 1960s. Farmers and then entrepreneurs devised ways to pump the water out and to spray it, ultimately developing center-pivot irrigation that has made regions with it very productive. But in the fifty years since this discovery, the aquifer is clearly undergoing depletion. At first the aquifer was estimated to contain more than three billion acre-feet, but it has lost hundreds of millions of acre-feet every year since the 1960s. Farmers have to pump the water from deeper and deeper wells, some parts of the aquifer are already depleted, and it recharges very slowly. Meanwhile, farmers in regions outside the aquifer's reach are still using dry farming techniques, and farmers in the region may have to return to those methods in a generation or two if the aquifer becomes too depleted to use (Little, 2009).

The US case resembles the Soviet one in that farmers in the Great Plains after World War I extended their farming in a gamble to survive falling prices, just as Russian and Soviet farmers extended their farming in a gamble to produce subsistence at least on some lands. In both cases, the gamble failed spectacularly in certain periods; in the Soviet case, resulting in major famines and substantial loss of life. Scientists and specialists in both countries studied what went wrong and came up with at least short-term solutions. For the Soviets, the solution was the Virgin Lands, which made overplanting and gambling that some crops would survive a core part of their farming system. For the United States, the solution was aquifer-based irrigation, which has been extremely successful but is clearly going to end. When it does it will leave the farmers in conditions similar to those of the 1930s, albeit with more highly advanced agricultural science to rely on.

But the US example applies to many other countries that have adopted similar irrigation systems and tapped their aquifers to help them farm arid regions. Most of these countries are exploiting aquifers that may never recharge, as in Saudi Arabia. Another case is Australia, which is an extremely productive agricultural country and has the Great Artesian Basin, an aquifer about four times the size of the Ogallala aquifer, in the western half of the country (Department of the Environment, Australian Government, n.d.). Yet Australia is more arid than the United States or even Russia, with chronic eight- to ten-year droughts, and its soils are very saline. By tapping aquifers, countries such as these are taking advantage of their place and their local underground resources, with a boom that will last for decades, but with the long-term prospect of a major crisis when the waters are depleted.

Contemporary Examples

Finally, some impression of what can happen when modern technology fails to consider the ecological limits of a particular place can be seen from three notorious but instructive development projects: the Tanganyika Groundnut Scheme from the last days of the British Empire, the Jari Project attempted by billionaire Daniel Ludwig in Brazil, and the depletion of the Aral Sea in the last decades of the Soviet Union. These projects attempted to modernize agricultural production (although not food production in the Jari project), and failed because they transgressed the limitations of place.

The Tanganyika Groundnut Scheme was a series of modernizing reforms introduced by the British and French governments to retain their empires, which some scholars have called the "second colonial occupation." The colonial powers had, by this time, amassed considerable scientific information about the colonies and, in addition, benefited from the scientific advances made before and during World War II. By 1955, British authorities had begun around seventy development projects, ranging from increasing food production to water and soil conservation (Hodge, 2007).

One of the pioneering projects was a 1947 plan to cultivate over three million acres in Tanganyika with peanuts (locally called groundnuts because of their growth pattern) to produce oil and other products for British needs and for sale in world markets. The plan included extensive mechanization of farming, rapid expansion of cultivation, construction of a railroad, a port, and other components, at an estimated cost of £24 million (worth about ten times that in 2014). British colonial specialists were divided about the feasibility of the project, but the prospect of modernizing farming and high production persuaded the government to authorize it.

The implementation of the project resembled a comedy where almost everything went wrong. Even though the planners had studied the region, they did not fully understand its specific characteristics. The tractors brought to Tanganyika to uproot the local trees and plow ended up frequently breaking down because the tree roots were too long and strong to remove easily. When finally some areas were cleared and ready to plow, the tractors turned out to be too heavy and damaged the soil, yet they were not powerful enough to cultivate the extremely hard soil in the dry season. By 1949, two years into the project, despite efforts to push the project forward rapidly, only fifty thousand acres, or 16 percent of the planned area, had been cultivated. By 1950, the British Parliament, extremely skeptical, agreed to alter the project into a more general development program. An investigation in September, however, found that the project had already cost six times as much as the value of the crops it produced, and by January 1951, the government abandoned the project, writing it off at a cost of £36.5 million or 50 percent above the planned total cost (Cavendish, 2001; Hodge, 2007).

The Groundnut Scheme is usually considered an example of the failures of colonialism, but it also reflects two other influences that relate to patterns discussed in the above examples. The project had originally been proposed by the Unilever Corporation but was then taken over by the government with the goal of producing edible oils to alleviate postwar shortages and rationing. The official in charge was the minister of food, John Strachey—an ex-communist who was familiar with Soviet history and development methods. Reflecting these two influences, the Groundnut Scheme combined an urgent commercial orientation, to produce groundnuts to be processed and sold as expeditiously as possible, with an implementation process evocative of the worst Soviet agricultural practices, such as those applied in the Virgin Lands. This late-colonial project thus combined capitalist and socialist high-pressure tactics and neglect of the local environment. It was no surprise that the project was so utterly insensitive to place (Hodge, 2007, ch. 7).

But purely capitalist projects could also go utterly awry. Billionaire Daniel Ludwig anticipated that demand for paper would grow rapidly and decided to speculate on this in 1967 by buying 1.6 million acres in the Brazilian rain forest (through connections with the then military dictatorship that ruled Brazil), employing thousands of Brazilians to cut down the rain forest and plant a type of tree that was well suited to pulp manufacturing. Here again everything went awry. Initial attempts to clear the forest with machinery damaged the topsoil, so the project had to employ manual labor for this task, which increased costs, required building settlements for the workers, and took much longer. The newly planted trees became infested with insects and grew slower and smaller than planned. The project had to find new species of trees, such as eucalyptus. Ludwig expanded the project to grow rice, but the soil was unsuited to the crop and needed considerable fertilizer. Management of the project was so difficult that Ludwig had to replace managers more than once every year. Staff and workers increasingly became ill with malaria and other diseases; meanwhile, the Brazilian government became impatient with the project's delays and tax-exempt status. The project's numerous changes and additional expenses cost Ludwig a billion dollars when he had expected to invest half of that at most. Returns from the project reached only seventy million dollars. Finally, in 1982, Ludwig transferred the project and its debts to a group of Brazilian businessmen who scaled it down drastically (Watkins, n.d.).

The Aral Sea, located between Kazakhstan and Uzbekistan, was—in 1950—one of the four largest lakes in the world, fed by two large rivers, the Syr Darya and Amu Darya, which originate in the Tian Shan and Pamir mountains that divide Central Asia from China. Under Khrushchev, the originator of the Virgin Lands project, the Soviet government decided to divert part of the waters from these rivers through canals to irrigate farmland to grow food and cotton in Uzbekistan. Over the next thirty years the project did succeed in growing considerable amounts of crops—Uzbekistan in 1988 exported more cotton than any other country. But it succeeded at great cost, because only a small part of the canals were lined, and half or more of the water leaked or evaporated. The Aral Sea steadily lost water, until by 1998, it had only 40 percent of its previous area and 20 percent of its previous water volume. The Soviet irrigation plans anticipated that the Aral Sea would shrink but did not foresee the extremely dangerous effects of that shrinkage. The Aral Sea bed, as a result of Uzbek farmers' use of fertilizers and pesticides, accumulated the runoff of these chemicals. As the Aral Sea shrinks, the seabed is exposed and dries, leaving a toxic dust that desert winds blow all over Central Asia—causing many serious diseases and increasing the region's mortality rates as well as damaging the crops that the diverted water is used to grow. The Aral Sea's former fishing industry is destroyed. Now, decades after the policy began, post-Soviet governments in the region are working with international agencies and scientists to find ways to mitigate the effects of the pollution and to at least partly restore the Aral Sea (Kasperson, Kasperson, & Turner, 1995; Micklin, 2007).

Conclusions and Observations

The historical record contains many cases of agricultural disasters caused by the specific effects of broader environmental events on local farming systems as well as by farmers and political leaders ignoring local and long-term environmental risks in their desire for short-term economic gains. People in the private sector, whether small farmers or billionaires, often ignore risks. Imperial governments, whether capitalist or communist, have that same tendency, motivated by a type of arrogance, especially when leaders (like Khrushchev) are ill educated and ideologically biased. Yet the Chinese case shows that well-intentioned governments can cope with quite serious disasters.

At present, societies have accumulated an abundance of scientific knowledge and historical studies that provide ample warning about the risks posed to food and agriculture in particular places from environmental threats, economic policies, and business practices. After all the disasters discussed above and many others that occurred because political and private-sector actors ignored place, there really is no further excuse to ignore or claim ignorance about environmental threats and ill-advised policies in vulnerable regions. This is especially the case because disasters that affect particular places can now have far greater global consequences than before, because the globalized economy communicates the effects of local or regional disasters widely and rapidly, and because the food needs of seven billion people approach the limits of global food productivity and reserves. The droughts in northern grasslands in 1988 reduced world food reserves to fifty-four days, less than the sixty-day minimum established as necessary by the United Nations Food and Agricultural Organization, the top international organization dedicated to agricultural research and policy formation. While some environmental threats, such as climate change, are global in scale, their effects will still vary by region, which requires specific responses by local populations and governments.

Before the modern period, most people obtained their food supplies from their own home region or at most through trade from nearby regions; food and place, and place and disasters, were tightly connected. In the modern period people increasingly obtain their food from remote regions, and even farmers in developed countries buy most of their food at supermarkets. The food crises before the modern period resulted in part from the limitations of transportation, which made it difficult to obtain food relief during disasters. Now the situation is reversed: the food system operates and depends on extensive use of transportation. This usually prevents food crises of the premodern period from occurring. For example, when drought destroyed crops in the American Midwest in 1988, no one starved as peasants starved after drought-induced crop failures in medieval China or nineteenth-century Russia. The old connection between food and place is now greatly reduced and has been replaced by the link between place and money: we obtain food in a store built by a corporation, which calculates that it can make a profit selling the food in our town. The USDA recently prepared a Food Access Research Atlas that shows the location of "food deserts," where many types of food cannot be purchased because few food stores have been located in those places (USDA, 2014).

This new dependence on transport instead of place, however, has also created a new connection between food and place, at least in most of the developed world. Now the connection is between the consumer in one place, the places where the consumer's food is produced (which consists of many other places), and the transport link between them. This means that an environmental disaster in one place, or a crisis in the transportation system in one region, will affect people—consumers—in many other places. Where the old link between food and place was more-or-less regionally limited, the new link is interregional and can be international or even global, depending on the way a particular food item is produced and marketed. This new relationship does not apply everywhere: the chronic crises in Sudan are geographically limited because that country has little involvement with world markets and is surrounded by the highly impassible Sahara desert. But most of the world is becoming increasingly dependent on transported food.

This situation is highly risky and is likely to end. Transport depends on oil and coal, and these minerals are finite in the long term and polluting in the short term. They are also vulnerable to political disruption. The "Hirsch report" on peak oil (US Department of Energy, 2005) warned that preparatory measures need to be taken at least twenty years in advance, or major disruptions will follow. Given the enormous interdependence and interconnectedness of the entire world economy, especially in agriculture, governments and societies need to make similar long-term preparations for the effects of global environmental disasters, taking into account the specific characteristics of different regions.

Beyond such an effort to anticipate a fuel and transport crisis, however, societies also need local food sources because of the vulnerability of transport, and also because a society that has too few people with concrete knowledge of farm production is extremely vulnerable if the transport link is disrupted. Our superior knowledge of every aspect of the environment and agricultural production, compared to the medieval period and before, can enable us to restore the old food link as a means to achieve food security while reducing the risk of the type of food-place disaster that happened so often in the past.

References

Aberth, J. (2010). *From the brink of the apocalypse: Confronting famine, war, plague and death in the later Middle Ages.* New York, NY: Routledge.

Adams, R. (1997). *Ancient civilizations of the New World.* Boulder, CO: Westview Press.

Adelman, J. (1994). *Frontier development: Land, labour, and capital on the wheatlands of Argentina and Canada, 1890–1914.* Oxford: Clarendon Press.

Bautier, R-H. (1971). *The economic development of Medieval Europe.* New York, NY: Harcourt Brace Jovanovich.

Blum, J. (1957). The rise of serfdom in Eastern Europe. *American Historical Review, 62*(4), 807–836.

Bray, F. (1984). *Science and civilisation in China, Vol. 6, Biology and Biological Technology; Part 2, Agriculture.* Cambridge: Cambridge University Press.

Cavendish, R. (2001, January 1). Britain abandons the groundnuts scheme, *History Today*, 51(1). Retrieved from http://www.historytoday.com/richard-cavendish/britain-abandons-groundnuts-scheme

Chen, Y., Syvitski, J. P., Gao, S., Overeem, I., & Kettner, A. J. (2012). Socio-economic impacts on flooding: A 4000-year history of the Yellow River, China. *Ambio, 41*(7), 682–698.

Coe, S. D. (1994) *America's first cuisines.* Austin: University of Texas Press.

Deng, G. (1999). *The premodern Chinese economy: Structural equilibrium and capitalist sterility.* New York, NY: Routledge.

Department of the Environment, Australian Government, *Great Artesian Basin*, (n.d.) Retrieved from http://www.environment.gov.au/topics/water/water-our-environment/great-artesian-basin

Diamond, J. (2005). *Collapse: How societies choose to fail or succeed.* New York, NY: Viking Press.

Duby, G. (1968). *Rural economy and country life in the medieval West.* Chapel Hill: University of North Carolina Press.

Elvin, M. (2004). *The retreat of the elephants: An environmental history of China*. New Haven, CT: Yale University Press.

Fagan, B. (2002). *The Little Ice Age: How climate made history, 1300–1850*. New York, NY: Basic Books.

Food and Agriculture Organization (FAO). (2014). Conservation Agriculture in Northern Kazakhstan. Food and Agricultural Organization of the United Nations, Technical Cooperation Programme. Retrieved from http://www.fao.org/tc/tcp/kazakhstan_en.asp

Hodge, J. M. (2007). *Triumph of the expert: Agrarian doctrines of development and the legacies of British colonialism*. Athens: Ohio University Press.

Hurt, R. D. (Ed.). (2002). *American agriculture: A brief history*. West Lafayette, IN: Purdue University Press.

Iannone, G. (Ed.). (2014). *The great Maya droughts in cultural context: Case studies in resilience and vulnerability*. Boulder: University Press of Colorado.

Jordan, W. C. (1996). *The great famine: Northern Europe in the early fourteenth century*. Princeton, NJ: Princeton University Press.

Kasperson, J., Kasperson, R., & Turner, B. L. (1995). *The Aral Sea Basin: A man-made environmental catastrophe*. Boston, MA: Kluwer.

Kates, S. (2013). *Defending the history of economic thought*. Cheltenham: Edward Elgar.

Kershaw, I. (1973). The Great Famine and agrarian crisis in England, 1315–1322. *Past and Present*, *59*(1), 3–50.

Li, L. M. (2007). *Fighting famine in North China: State, market, and environmental decline, 1690s–1990s*. Stanford, CA: Stanford University Press.

Little, J. B. (2009, March 1). The Ogallala Aquifer: Saving a vital U.S. water source. *Scientific American*. Retrieved from http://www.scientificamerican.com/article/the-ogallala-aquifer/

McCauley, M. (1976). *Khrushchev and the development of Soviet agriculture: The Virgin Lands Programme, 1953–1964*. New York, NY: Holmes and Meier.

Meinig, D. W. (1962). *On the margins of the good earth: The South Australian wheat frontier, 1869–1884*. Chicago, IL: Rand McNally.

Micklin, P. (2007). The Aral Sea disaster. *Annual Review of Earth and Planetary Sciences*, *35*(4), 47–72.

Monahan, W. G. (1993). *Year of sorrows: The Great Famine of 1709 in Lyon*. Columbus: Ohio State University Press.

Moon, D. (2013). *The plough that broke the steppes*. Cambridge: Cambridge University Press.

Opie, J. (1992). The drought of 1988, the global warming experiment, and its challenge to irrigation in the old Dust Bowl region. *Agricultural History*, *66*(2), 279–306.

Pringle, P. (2008). *The murder of Nikolai Vavilov*. New York, NY: Simon & Schuster.

Roll-Hansen, N. (2005). *The Lysenko effect: The politics of science.* Amherst, MA: Humanity Press.

Shabykina, L. V. (1976). *Iarovaia pshenitsa v severnom Kazakhstane.* Alma-Ata: izdatel'stvo "Kainar."

Stearns, P. (1998). Why study history? *American Historical Association.* Retrieved from http://historians.org/about-aha-and-membership/aha-history-and-archives/archives/why-study-history-%281998%29

Tauger, M. B. (2001). Grain crisis or famine? The Ukrainian state commission for aid to crop failure victims and the Ukrainian famine of 1928–1929. In D. J. Raleigh (Ed.), *Provincial landscapes: Local dimensions of Soviet power* (pp. 146–170). Pittsburgh, PA: University of Pittsburgh Press.

Tauger, M. B. (2006). Stalin, Soviet agriculture, and collectivization. In F. Trentmann & F. Just (Eds.), *Food and conflict in Europe in the age of the two World Wars* (pp. 109–142). New York, NY: Palgrave Macmillan.

Tauger, M. B. (2011a). *Agriculture in world history.* London: Routledge.

Tauger, M. B. (2011b). Famine in Russian history. *Supplement to the Modern Encyclopedia of Russian and Soviet History*, vol. 10. Gulf Breeze: Academic International Press, 79–92.

Tauger, M. B. (2012). Pavel Pantelimonovich Luk'ianenko and the origins of the Soviet Green Revolution. Forthcoming in a collection of articles from the Second International Workshop on Lysenkoism, held in Vienna, Austria, June 2012.

US Department of Energy (2005). *Peaking of world oil production: Impacts, mitigation and risk management* (Hirsch Report). Retrieved from http://www.netl.doe.gov/publications/others/pdf/oil_peaking_netl.pdf

US Department of Agriculture (USDA). (2014). *Food access research atlas.* Retrieved from http://www.ers.usda.gov/data-products/food-access-research-atlas.aspx#

Watkins, T. (n.d.), The Jari Project of Brazil. San Jose State University Department of Economics. Retrieved from http://www.sjsu.edu/faculty/watkins/jari.htm

Welton, G. (2011). *The impact of Russia's 2010 grain export ban.* Oxfam. Retrieved from http://www.oxfam.org/sites/www.oxfam.org/files/file_attachments/rr-impact-russias-grain-export-ban-280611-en_3.pdf

Will, P-E., & Wong, R. (1991). *Nourish the people: The state civilian granary system in China, 1650–1850.* Ann Arbor: Center for Chinese Studies, University of Michigan.

Worster, D. (1979). *The Dust Bowl: The Southern Plains in the 1930s.* Oxford: Oxford University Press.

Part II

Social and Cultural Contexts

Chapter 4

Ways of Knowing the World: The Role of Knowledge and Food Movements in the Food-Place Nexus

Jennifer Sumner

Introduction

Food and knowledge have been linked since our earliest ancestors gathered and hunted in the forests and grasslands where they first evolved. Which plants were good to eat, why game was sometimes plentiful, and how to safely access honey were just a few of the endless questions concerning food that knowledge provided answers for. Then, as now, such knowledge was grounded in place—situated in the locations where food was found and eaten.

Over time, knowledge about food became codified into ways of knowing—epistemological perspectives that focused on certain kinds of knowledge. Although there are many ways of knowing, this chapter will concentrate on two—traditional knowledge and expert knowledge—and their relationship to food and place. It will begin by discussing traditional knowledge before moving on to trace the rise of expert knowledge in the age of neoliberalism as it seeks to invalidate traditional knowledge and literally dis-place us from the food we eat. The chapter will then examine social movements and the hybrid knowledge associated with food movements, that is, knowledge that is situated in the food-place nexus and draws both from traditional knowledge and from expert knowledge to create novel solutions to what has come to be known as the global food crisis. Associated with inflated food prices, this crisis has emerged from a number of factors: industrial capitalism's long-term dependence on fossil fuels, the inflationary effects both of biofuel offsets and of financial speculation, and the policies of the corporate food regime that promote the concentration and centralization of agribusiness capital (McMichael, 2009).

Traditional Knowledge

Knowledge has been defined as "the way people understand the world, the way in which they interpret and apply meaning to their experiences" (IAASTD, 2009, p. 285). As such, knowledge is a social artifact that is centered on collective meaning. Being a social artifact, however, makes knowledge endlessly complex. It can be used for good or evil, which means that its moral and political implications must always be considered (Seidman in Solomon, 2005).

Like food, knowledge is both produced and consumed, and the production and consumption of food and knowledge have mutually informed each other for millennia. In myriad ways, producing and consuming food has generated a great deal of knowledge that in turn has been processed, shared, reflected upon, and applied in the further production and consumption of food. Within this dialectical relationship, place has always figured strongly, leaving an indelible mark on this vital interface of human engagement.

Over time, knowledge about food that had accumulated through trial and error solidified into what is broadly known as traditional knowledge, understood as "a cumulative body of knowledge, know-how, practices and representations maintained and developed by peoples with extended histories of interaction with the natural environment" (International Council for Science, 2002, p. 3). Such knowledge is predominantly place-based, emerging from the interaction between humans and the food they produce and consume in the locales where they live. Traditional knowledge is foundational to local level decision making about many basic aspects of life, including hunting, fishing, gathering, and agriculture, as well as the preparation, conservation, and distribution of food (International Council for Science, 2002).

As an evolving accumulation of practices and beliefs passed down through the generations, traditional knowledge is related to, but not synonymous with, other forms of place-based knowledge, such as indigenous knowledge and local knowledge. It differs from indigenous knowledge in that the latter emphasizes length of time in one location, as evidenced by Dei, Hall, and Rosenberg (2000), who associate indigenous knowledge with the long-term occupancy of a certain place. It also differs from local knowledge, which is more connected to a given culture or society (IAASTD, 2009), whereas traditional knowledge may not be so narrowly constituted.

In common with these other forms of place-based knowledge, traditional knowledge is a kind of "situated" knowledge—embedded in a physical site or location and remaining relevant over time (Sole & Edmondson, 2002). Examples include traditional knowledge about local plant varieties, such as types of rice in India that are known to be tolerant of drought or heat, or the dozens of varieties of potatoes in the Andes that are adapted to all sorts of mountainous terrain. Another example is seasonality, which celebrates local fruits and vegetables in season. A final example is *terroir*—the notion that place bestows special characteristics on food (Metcalf Foundation, 2008), which is important in sectors such as the wine industry. The salience of place and traditional knowledge about food, however, is being obscured by the restless placelessness of a global food regime that is driven by neoliberal policy and abetted by another form of knowledge–expert knowledge.

Neoliberalism and Expert Knowledge

Knowledge, like food, is always produced and consumed within a larger context. For several decades, the set of interrelated conditions that has influenced the production and consumption of knowledge has been neoliberalism, which Brenner, Peck, and Theodore (2010) maintain, involves extending market-based competition and processes of commodification into areas of life that were previously insulated from these influences. Geographer David Harvey (2006) has defined neoliberalism as "a theory of political economic practices which proposes that human well-being can best be advanced by the maximization of entrepreneurial freedoms within an institutional framework characterized by private property rights, individual liberty, free markets, and free trade. The role of the state is to create and preserve an institutional framework appropriate to such practices" (p. 145). To this definition, Harvey (2005) adds that if markets do not already exist (for example, in education or water), then they must be created, by state action if required. But beyond creating the framework for maximizing entrepreneurial freedoms and the markets entrepreneurs' use, he maintains, the state has no role other than to enforce the proper functioning of these markets (e.g., through the police or the military). In contrast to the strong role of the state under Keynesianism, state intervention under neoliberalism becomes minimal because the theory posits that "the state cannot possibly possess enough information to second-guess market signals (prices) and because powerful interest groups will inevitably distort and bias state interventions (particularly in democracies) for their own benefit" (Harvey, 2005, p. 2).

In effect, the ongoing process of neoliberalization involves three approaches: it prioritizes market responses to regulatory problems; it intensifies the commodification of all aspects of social life; and it often uses speculative financial instruments to pry open new areas for profit-making (Brenner et al., 2010). By seeking to bring all human action into the domain of the market, the process of neoliberalization "has entailed much 'creative destruction,' not only of prior institutional frameworks and powers (even challenging traditional forms of state sovereignty) but also of divisions of labour, social relations, welfare provisions, technological mixes, ways of life and thought, reproductive activities, attachments to the land and habits of the heart" (Harvey, 2005, p. 3).

The power and reach of neoliberalism have grown since it first gained prominence in the 1970s and was taken up by Ronald Reagan and Margaret Thatcher in the 1980s. Strengthened by the fall of the Soviet Union in 1989, it became globally entrenched through free-trade agreements and the creation of the World Trade Organization (WTO) in 1995, both of which became vehicles for embedding rules limiting government powers to intervene in the economy (Roberts, 2013). Neoliberalism's focus on entrepreneurialism, market exchange, and commodification has had a profound influence on all areas of existence, including food, place, and knowledge.

In terms of food, Guthman (2008) maintains that neoliberal policies have fostered the restructuring of agriculture and food sectors through the privatization of land and water rights, the use of free-trade agreements to dismantle national-level food safety regulations, and the protracted dismantling of food-oriented entitlement programs set up to combat hunger. The global food crisis is a crushing indictment of these policies, ushering in record levels of hunger for the world's poor, while at the same time producing record profits for the world's major agri-food corporations (Holt-Giménez, 2009; Holt-Giménez & Shattuck, 2011). Neoliberalism has become so pervasive that it has even had a lasting influence on attempts to oppose these policies in the food and agricultural sectors. In other words, neoliberalism has also shaped the "politics of the possible" (Guthman, 2008, p. 1180) in that efforts to develop alternatives and/or to redress some of the negative implications of neoliberal policies are still framed within its worldview, thus continuing to condition our relationship to food and significantly narrowing how we imagine the future.

For place, neoliberalism has in many ways erased the notion of place through its global ideology. The defense of place is pejoratively labeled "protectionism." No longer bound to place, workers have become increasingly mobile—uprooting themselves in search of employment, precarious or otherwise. Entrepreneurs roam the planet, searching for profit opportunities and never committing to one place. And food is increasingly placeless—grown or raised as cheaply as possible in one location, processed in another, packaged in a third, and sold anywhere in the world that people can afford it. All these outcomes diminish the importance of place as "a topographically and ecologically situated, inhabited space, a locality whose boundaries are porous, but even so, a particular world, with its own historical memory and shared understanding of itself" (Dirlik, 1999, in Friedmann, 2002, p. 311).

For knowledge, the rise of neoliberalism has challenged traditional knowledge through its drive to commodify all aspects of human life for private profit. Along with genes, body parts, and death itself, knowledge has become a product that is bought and sold on the global market. To gain commodity status, knowledge had to be privatized, packaged, and sold for a profit. Knowledge that does not fit easily into this process—such as the management-intensive (rather than commodity-intensive) knowledge associated with organic agriculture—is marginalized, sidelined, ridiculed, or vilified. In other words, if knowledge does not provide an opportunity to maximize entrepreneurial freedoms, it is worthless in the eyes of the market.

It is not surprising then that in this age of neoliberalism, we are living in what has been dubbed the knowledge economy—the idea that the economy is dependent on knowledge as input and output (Gibb & Walker, 2013). Intellectual property rights, university research parks, and proprietary knowledge are all evidence of this phenomenon. But given that economies are based on the production and distribution of goods and services in short supply, and that there has never been a shortage of knowledge, the concept is an oxymoron (Sumner, 2006)—a contradiction in terms that is used to justify the commodification of knowledge.

To legitimize this commodification process and to delegitimize traditional knowledge, the product promoted by the global market is referred to as "expert" knowledge, also known as scientific or managerial knowledge (Fonte, 2008). McBride and Burgman (2012) characterize expert knowledge as the knowledge that qualified individuals have gained through their technical practices, training, and experience. This knowledge can include "recalled facts or evidence, inferences made by the expert on the basis of 'hard facts' in response to new or undocumented situations, and integration of disparate sources in conceptual models to address system-level issues" (p. 13). Although expert knowledge has long existed in some form, it has gained tremendous currency under neoliberalism. But to compete with something that is freely shared and given—traditional knowledge—those engaged in the so-called knowledge economy have belittled and marginalized it, while touting the kind of knowledge that can be privatized, packaged, and sold to bolster corporate profits. For example, the Green Revolution offered farmers in the developing world the expert knowledge of industrial agriculture, which is capital-intensive because it substitutes machinery and purchased inputs for the labor of humans and animals (IAASTD, 2009). The Green Revolution displaced traditional forms of agricultural knowledge, while opening opportunities for capital accumulation at every step of the global food chain by creating dependency on purchased inputs.

Although vital to the human enterprise, expert knowledge has become a kind of official knowledge that supports the dominant groups in society. Cunningham (1988) explained the role of such knowledge: "All bodies of knowledge are not equal. The knowledge that supports the dominant paradigms in a culture is systematically produced, disseminated, and reified so as to become common sense to the average citizen" (p. 137). The role of official knowledge highlights issues of power associated with different bodies of knowledge.

Knowledge and Power

The philosopher Bertrand Russell (1967) defined power as the production of intended effects. With this in mind, we can understand, as Stone (2000) does, that knowledge equals power because it can produce intended effects in many areas of life, including food. For Stone, knowledge informs, enables, and empowers not only those who possess it but also the institutions that embody it. Given this association with power, any reference to knowledge "implies a struggle between different 'knowledges' or what are often described as 'discourses,' 'world views' and 'regimes of truth'" (p. 247).

Struggle is evident in Wolfgang Sachs's (2007) understanding of the power of knowledge to direct people's attention by highlighting a certain reality, while ignoring other ways of relating to the world. This helps to explain the ability of expert knowledge to obscure other ways of knowing, including those associated with traditional knowledge. The loss of traditional knowledge around seed saving, plant varieties, integrated pest management, animal traction, canning, pickling, and root cellars testifies to the "asymmetries of power" (Fonte, 2008, p. 202) that allow expert knowledge to block out the accumulated understanding of centuries of food practices and beliefs. But the very dynamics that valorize expert knowledge throughout the food chain have also enabled traditional knowledge to make a comeback by spawning resistance in the form of food-related social movements.

Social Movements, Food, Knowledge, and Place

The shift from traditional to expert knowledge in the age of neoliberalism not only highlights the power issues surrounding knowledge, but also signifies a broader shift in ruling relations from the public to the private sector. By prioritizing private property, neoliberalism promotes the privatization of formerly public resources, such as water, healthcare, hydroelectricity, and postal systems. In response to this power shift, a number of social movements have arisen—the recent Occupy Movement, the antiglobalization movement at the turn of the millennium, and the postautistic economics movement are just some of the forms of protest against the priorities of neoliberalism. All the chapters of these movements are grounded in a place, but are linked globally through technologies such as social media. In terms of food, opposition to the negative effects of neoliberalism is coming from a range of food movements, such as the local food movement, the organic food movement, the Slow Food movement, the fair trade movement, and the food justice movement. Although these food movements may appeal to different groups of people with different interests, they are ultimately grounded in the places where members of these movements carry out their various forms of resistance: buying local food, farming organically, holding slow-food dinners, serving fair-trade products, or handing out free food to the homeless.

Social Movements

Morris (2005) describes social movements as "a wide variety of collective attempts to bring about a change in social institutions or to create a new social order" (p. 589). While not all social movements can be regarded as positive (e.g., the white supremacist movement), most work toward progressive social change, whether they deal with labor, peace, the environment, women, or civil rights. In this way, social movements can act as "the lightning rods of society" (Walters, 2005, p. 139) and often represent a challenge to the status quo.

Social movements are sites both of knowledge production and of consumption. In terms of the latter, participants search out the knowledge they require to begin the process of effecting social change. Miles (1996) observes that people engaged in social struggle make decisions about what and why they need to learn, not only learning from each other, but also from other educational resources of their choosing. In terms of the former, social movements produce knowledge that others can learn from. Insights regarding the effects of patriarchy, the problems of industrial agriculture, and the injustices of systemic racism represent just some of the vast wealth of knowledge produced by social movements. Put another way, social movements are both "*bearers of knowledge* on a range of issues such as injustice, politics, grievances, social divisions and power structures and the power of science and technology" and "*producers of knowledge*"—involved in knowledge generation (Flowers & Swan, 2011, p. 236).

The sheer quantity of knowledge produced and consumed by social movements has given rise to the term *social movement learning*. Hall and Clover (2005) understand *social movement learning* as learning by people who are either part of a social movement or are outside it and learn from the actions or the existence of social movements. For example, Hall (2006) points out that many men are not members of the women's movement, but they have learned a great deal about the fair treatment of women from the movement. In this way, social movements make power visible by challenging the status quo through marches, slogans, demonstrations, rallies, sit-ins, and occupations. Moreover, successful social movements can actually change how we engage with knowledge and the relations of power. Finger (1989, in Hall, 2006) goes so far as to argue that social movement learning is more powerful than all the learning taking place in schools. This is certainly apparent in the realm of food.

Food Movements

A number of social movements have coalesced around the production and consumption of food. These food movements involve "a whole range of political practices around organic foods, local foods, anti-industrialised food production, vegetarianism, food for social justice, Slow Food movement, food security, to name but a few," all with the aim of getting us to change what and how we eat (Flowers & Swan, 2011, p. 235). As a reaction to the power shifts engendered by neoliberalism, these food movements recognize that the production of industrial food needs reform because its costs, in terms of society, the environment, public health, and animal welfare are too high (Pollan, 2010).

Like other social movements, food movements work toward positive social change, including more environmentally friendly growing conditions, greater protection of artisanal foodways, fairer prices for producers, and practical solutions to the hunger/obesity crises. Their fairly recent emergence mirrors the growing contradictions of neoliberal capitalism in which maximizing profits trumps basic human needs in the food system (Winson, 2010). Flowers and Swan (2011) argue that the politics of food knowledge is central to many of these food movements: "The politics of knowing, what is known, who produces it and 'who is in the know' are critical to food social movements" (p. 236). They use the term *food knowledges* to depict the sheer range of understandings and practices surrounding food, including: "knowledge which is codified and informal; academic and experiential; lay and expert; embodied and cognitive; individual and institutional; emotional and logical; everyday and rarefied; gendered, racialised and classed" (p. 235).

The idea of food knowledge brings to mind the work of German philosopher Jürgen Habermas (1978), who put forward the idea that humans have an interest in three kinds of knowledge. The first kind he called empirical/analytic knowledge, which Morrow and Torres (1995) describe as being "based upon a desire potentially to control through the analysis of objective determinants" (p. 24). Expert knowledge fits into this category: how to repair a tractor, how to use a microwave oven, and how to grow tomatoes are all examples of empirical/analytic knowledge in the realm of food, as are questions of how to build a root cellar, how to bake a cake from scratch, and how to can peaches. The second kind of knowledge is referred to as historical-hermeneutic knowledge. This is knowledge that Morrow and Torres (1995) describe as based on the desire to "understand through the interpretation of meanings" (p. 24). Discourse analysis of food ads, narratives about comfort food, and discussions of women's complex relationship with food are all forms of historical-hermeneutic knowledge. The third kind of knowledge is called critical-emancipatory knowledge. This involves knowledge "based upon a desire potentially to . . . transform reality through the demystification of falsifying forms of consciousness" (Morrow & Torres, 1995, p. 24). The transformative learning of conventional farmers who transition to organic production, critical food literacy, and guerrilla gardening are all associated with critical-emancipatory knowledge. Although the first kind of knowledge has gained the status of "official knowledge," Habermas (1978) has argued that all three are vital to the human enterprise.

By championing all three kinds of knowledge, food movements have eschewed the traditional/expert knowledge divide in their struggle to resist the dictates of the neoliberalized food system. They have not only revalidated the traditional knowledge associated with the production and consumption of food in a particular place (and thus reclaimed place as central to food production and consumption) but have also adopted the positive aspects of expert knowledge and used it to create new forms of knowledge. In this way, they have become vibrant sites of hybrid knowledge production, offering an instructive template for considering how to craft solutions to the global food crisis. Three examples will illustrate the power of new forms of knowledge in social movements: the fair trade movement, La Vía Campesina, and the local food movement.

The Fair Trade Movement

Based on the principle of fair pricing, not market pricing, fair trade involves "a trading partnership, based on dialogue, transparency, and respect that seeks better trading conditions for, and securing the rights of, marginalized producers and workers, especially in the South" (EFTA, 2001). Jaffee, Kloppenburg, and Monroy (2004) see it as an alternative market system with the aim of redressing the historically inequitable terms of trade between the North and South and forging more direct linkages between producers and consumers. In a world dominated by so-called "free trade" (although some argue that there is no such thing as free trade—see Galbraith, 2008; Hamilton, 2001; Norman, 2013), fair trade raises questions of justice, equity, and transparency.

The fair trade movement emerged in the 1960s as a response to the problems associated with increasing globalization. In the beginning, the movement consisted of a small group of pioneers who concentrated on the practicalities of trade done very differently, investigating how trade could be made to work for, not against, poor producers in the Global South (Ransom, 2001). From these humble origins, fair trade reached sales of $6 billion by 2011 (Roberts, 2013) and is still growing. The movement deals in a number of products, including clothing, flowers, and food items such as coffee, tea, chocolate, bananas, and spices. By inducing customers in the Global North to pay a premium for these products, it ensures a more full-cost accounting than is currently available under neoliberalism, which downloads many of the costs of producing food onto society and the environment (Sumner, 2008).

The knowledge that the fair trade movement offers the world is embodied in the concept of fair trade itself. A Trojan-horse term, it implies that other forms of trade are not fair and therefore invites scrutiny of them. According to Ransom (2001), this scrutiny focuses on examining who makes the things we consume and how they are made. Whereas other forms of trade rely on consumers remaining blind to such discoveries, fair trade "encourages us to take a closer look, to engage more critically with the intriguing, sometimes shameful, details of everyday human life" (p. 22).

One of the great strengths of fair trade is that it operates both inside the market and against it (Raynolds, 2002), thus generating knowledge of actual working alternatives to conventional trade and the neoliberal market. In doing so, it promotes social and environmental goals, as well as economic ones. For example, "Generalizing across commodities, at a minimum, fair trade standards are enacted by a price premium, a guaranteed price floor, long-term trading contracts, easier access to credit, and shorter supply chains. In turn, the cooperatives growing these products must be democratically organized and utilize the fair trade premium for the benefit of members. Also, producers commit themselves to improving the environmental conditions of production by reducing or avoiding pesticide use" (Goodman, 2004, p. 897).

One of the weaknesses of the fair trade movement is that most organizations involved in fair trade are only mandated to deal with farmers in the Global South, leaving farmers in the Global North without the benefits of fair trade. This is beginning to change, however, with the application of fair-trade principles to farmers closer to home. For example, Farmer Direct, a cooperative of seventy certified organic farms in Saskatchewan, is the first business in North America to receive domestic fair-trade certification.

Evidence of the effectiveness of the knowledge produced by the fair trade movement can be found in the establishment of fair-trade towns, which promote fair trade as a community effort. The first fair-trade town was declared in Garstang, England, in 1999, and spread to North America, when Wolfville, Nova Scotia, became the first Canadian fair-trade town in 2007 (Fairtrade Canada, 2012). To become a fair-trade town, communities must achieve six goals:

1. The local council uses fair-trade-certified products and supports the fair-trade towns campaign
2. Stores and restaurants serve fair-trade-certified products
3. Workplaces, faith groups, and schools use and promote fair-trade-certified products
4. Public awareness events and media coverage are held on fair trade and the campaign
5. A steering group is created for continued commitment
6. Other ethical and sustainable initiatives are promoted within the community (Fairtrade Canada, 2012).

In this way, the knowledge production associated with the fair trade movement spans the world, linking producers and consumers in different places through a novel solution to the global food crisis.

La Vía Campesina

La Vía Campesina is an international peasant association that brings together 148 organizations based in 69 countries in the Americas, Europe, Asia, and Africa (Desmarais, 2012). As the world's largest social movement (Friedmann, 2012), La Vía Campesina has critiqued the privatization of public rights and resources advocated by the neoliberal agenda of the WTO. A place-based food movement with international reach, La Vía Campesina is rooted in the land because, first and foremost, campesinos come from the countryside (Desmarais, 2007). Focusing on the needs of the world majority, La Vía Campesina has argued that "the food crisis—now linked to the economic and environmental crisis—is the direct result of decades of destructive policies that spurred the globalization of a neoliberal industrial and corporate-led model of agriculture" (Desmarais, 2012, p. 359).

The new form of knowledge that La Vía Campesina has offered to the world as a solution to the food crisis can be condensed into the innovative concept of food sovereignty. Born out of struggles for power in the global food system, food sovereignty offers a vision to redress the abuse of the powerless by the powerful throughout the food system (Patel, 2007). In this vision, food sovereignty promotes "the right of peoples and nations to control their own food and agricultural systems, including their own markets, production modes, food cultures, and environments" (Wittman, Desmarais, & Wiebe, 2010, p. 2). Unlike food security, which involves *access* to food, food sovereignty advocates for the *right* to food and to a just livelihood for those who work in the food chain, from field to table historically, the people who are the most impoverished, the most exploited, and the most exposed to hazardous chemicals (Albritton, 2009). It also highlights the ongoing consolidation of power in the hands of a few multinationals that aim to control the global food system and thus increasingly control what we eat. For example, Hauter (2012) reports that twenty food corporations produce the majority of food consumed by Americans, including organic brands. Patel (2007) informs us that six companies control 70 percent of wheat trade, one controls 98 percent of packaged tea, and 20 control the world coffee trade. And Goodall (2005) points out that ten multinational corporations control over half the world's food supply. In particular, food sovereignty offers a vision of alternative food systems based on consensus—not competition—that can "feed the world and cool the planet" (La Vía Campesina, 2010, in Desmarais, 2012, p. 372).

La Vía Campesina has, in turn, sparked an indigenous food-sovereignty movement, based on indigenous people's knowledge of their distinct cultures and relationships to place and to food systems in their traditional territories. Like food sovereignty itself, indigenous food sovereignty "provides a framework for exploring, transforming, and rebuilding the industrial food system towards a more just and ecological model for all" (Morrison, 2011, p. 98). In essence, indigenous food sovereignty reaches across cultures to highlight the ways that food-related action and policy reform can be infused with indigenous knowledge, values, wisdom, and practices, offering guidance in "changing human behaviour and ending destructive relationships to Mother Earth and the land and food systems that sustain all human beings" (p. 112). In this way, the knowledge production associated with La Vía Campesina negates Margaret Thatcher's infamous statement that there is no alternative to neoliberalism, by presenting a novel solution to the global food crisis that is rooted in place and linked globally.

The Local Food Movement

As another social movement that has arisen due to the power shifts associated with neoliberalism, the local food movement has developed into a veritable knowledge hub. Clearly place sensitive, as well as being values oriented and participatory in nature (DeLind, 2011), the local food movement has been described as: "A collaborative effort to build more locally based, self-reliant food economies—one in which sustainable food production, processing, distribution, and consumption [are] integrated to enhance the economic, environmental, and local health of a particular place" (Feenstra, 2002, p 100).

Like any social movement, the local food movement exhibits both strengths and weaknesses, but it has indubitably changed our relationship with food through the new forms of knowledge associated with it. A few concepts will illustrate this point: food miles, the 100-mile diet, locavores, and foodsheds. *Food miles* describes the distance that food travels from field to fork and raises the issue of the oil and carbon emissions embedded in what Gussow (2001) has described as "jet-lagged food" (p. 222). Although the concept can divert from other important issues, such as social justice and sustainable forms of agriculture, it has managed to focus people's attention on the place their food comes from (or on their inability to locate where it comes from). It has also highlighted the irrationality of what Pollan (2006) refers to as "redundant trade" (p. 45), which involves the simultaneous and often superfluous import and export of the same food items (for example, California imports as many cherries and almonds as it exports).

An associated concept that has captured the public imagination is the *100-mile diet.* Championed by Alisa Smith and J. B. MacKinnon's (2007) description of living for a year on purely local food, it was simultaneously popularized by novelist Barbara Kingsolver (2007) in her investigative memoir, *Animal, Vegetable, Miracle: A Year of Food Life.* The concept has, in turn, spawned 100-mile markets—stores dedicated to selling only goods that come from within a one-hundred-mile radius. Both concepts help to relink producers and consumers and to reconstruct the lost infrastructure of a local food system, which is crucial in the neoliberal age of long supply lines that can get interrupted by environmental or political events.

Locavore is yet another concept to emerge from the local food movement. It means "a person who prefers to eat (or only eats) from within his/her own region" (DeLind, 2011). Voted "word of the year" in 2007 by the New Oxford American Dictionary, the term was coined two years earlier by a group of women in San Francisco, who proposed that local residents should try to eat only food that was grown or produced within a hundred miles of the city (OUP, 2007). In essence, "it captured the ethos of those who realized that the price of supermarket food did not reflect its true cost" (Elton, 2010) and encouraged consumers to consider place when choosing food. The concept has helped to alter consumer preference and behavior—one of the vital aspects of food regime change.

A final, though less well-known, example of the new forms of knowledge in the local food movement is the concept of a *foodshed*, which describes how food flows into and out of a particular place.

First developed in 1929, the term was reintroduced by Getz (1991), who described it as the area defined by a structure of supply. The term was subsequently popularized by Kloppenburg, Hendrickson, and Stevenson (1996, p. 34), who saw it as a "unifying and organizing metaphor for conceptual development that starts from a premise of the unity of place and people, of nature and society." For Kloppenburg et al., *foodshed* is a bridging term, connecting thinking and doing, theory and action—all leading to the type of analysis that can foster food-system change. It is also a vehicle through which people can reassemble their fragmented identities, reestablish community, and "become native, not only to a place but to each other" (p. 34). In this way, the knowledge production associated with the local food movement comes to permeate everyday thought, using place as a springboard for novel solutions to the global food crisis.

Conclusions and Observations

This chapter has provided an overview of the role of knowledge and food movements in the food/place nexus. Although there are many kinds of knowledge, the chapter has focused on two—traditional knowledge and expert knowledge—and the part played by neoliberalism in promoting the latter, while edging out the former. It has also highlighted the relationship between knowledge and power and the struggle to ensure that certain kinds of knowledge become and remain dominant. As vibrant sites of knowledge production and consumption, food movements overcome this struggle by utilizing both forms of knowledge. Three food movements were discussed in terms of this hybrid knowledge: the fair trade movement, La Vía Campesina, and the local food movement.

Food movements typify what Eyerman and Jamison (1991 in Walters, 2005, p. 140) noted about social movements and new forms of knowledge: "Social movements are not merely social dramas; they are the social action from where new knowledge including worldviews, ideologies, religions, and scientific theories originate." Anchored in place, food movements produce forms of knowledge that affect our everyday lives. They are changing our outlook on food by encouraging us to think more deeply about place and the origins of our food. They are also promoting new ways of knowing by combining traditional knowledge and expert knowledge in their quest for change and their reaffirmation of place in a globalizing world. In addition, these movements are expressing reverence for Mother Earth, reminding us that we are inextricably linked to place, not only physically, but also spiritually and culturally. And, finally, they are generating new concepts and theories to help us frame the way we understand the food/place nexus. In this way, food movements provide the springboard for innovative solutions to the global food crisis, using place as the starting point.

Maturana and Varela (1987) have described knowledge as "effective action in the domain of existence" (p. 244). By this description, food movements can be understood as influential knowledge producers that can help us in the broadest sense to "know our place" when we eat. In addition, their local embeddedness, global reach, broad appeal, and ability to form coalitions (e.g., organic fair-trade food) speak to a future that honors place, instead of erasing it; that develops cooperative fair-trade agreements, instead of competitive free-trade agreements; that cares for the land by practicing organic agriculture, instead of industrial monocropping; that taxes junk foods, instead of subsidizing them; that supports science for the public good, not science for private capital accumulation; and that ensures that everyone is fed, without harming the environment.

References

Albritton, R. (2009). *Let them eat junk: How capitalism creates hunger and obesity*. Winnipeg: Arbeiter Ring Publishing.

Brenner, N., Peck, J., & Theodore, N. (2010). After neoliberalization? *Globalizations, 7*(3), 327–345.

Cunningham, P. M. (1988). The adult educator and social responsibility. In R. G. Brockett (Ed.), *Ethical Issues in Adult Education* (pp. 133–145). New York, NY: Teachers College Press.

Dei, G. J. S., Hall, B. L., & Rosenberg, D. G. (Eds.). (2000). *Indigenous knowledges in global contexts: Multiple readings of our world*. Toronto: University of Toronto Press.

DeLind, L. B. (2011). Are local food and the local food movement taking us where we want to go? Or are we hitching our wagons to the wrong stars? *Agriculture and Human Values, 28*(2), 273–283.

Desmarais, A. A. (2007). *La Vía Campesina: Globalization and the power of peasants*. Halifax: Fernwood Publishing.

Desmarais, A. A. (2012). Building food sovereignty: A radical framework for alternative food systems. In M. Koç, J. Sumner, & A. Winson (Eds.), *Critical perspectives in food studies* (pp. 359–377). Toronto: Oxford University Press.

European Fair Trade Association (EFTA). (2001). Fair trade in Europe 2001. Maastricht: European Fair Trade Association.

Elton, S. (2010). *Locavore: From farmers' fields to rooftop gardens—How Canadians are changing the way we eat*. Toronto: HarperCollins.

Fair Trade Canada. (2012). Fair trade towns. Retrieved from http://fair-trade.ca/en/get-involved/fair-trade-towns

Feenstra, G. (2002). Creating space for sustainable food systems: Lessons from the field. *Agriculture and Human Values, 19*(2), 99–106.

Flowers, R., & Swan, E. (2011). 'Eating at us': Representations of knowledge in the activist documentary film *Food, Inc. Studies in the Education of Adults, 43*(2), 234–250.

Fonte, M. (2008). Knowledge, food and place: A way of producing, a way of knowing. *Sociologia Ruralis, 48*(3), 200–222.

Friedmann, H. (2012). Changing food systems from top to bottom: Political economy and social movements perspectives. In M. Koç, J. Sumner, & A. Winson (Eds.), *Critical perspectives in food studies* (pp. 16–32). Toronto: Oxford University Press.

Friedmann, J. (2002). Placemaking as project?: Habitus and migration in transnational cities. In J. Hillier & E. Rooksby (Eds.), *Habitus: A sense of place* (pp. 299–316). Burlington, VT: Ashgate Publishing.

Galbraith, J. K. (2008). *The predator state: How conservatives abandoned the free market and why liberals should too.* New York, NY: Simon & Schuster.

Getz, A. (1991). Urban foodsheds. *The Permaculture Activist, 24* (October), 26–27.

Gibb, T., & Walker, J. (2013). Knowledge economy discourses and adult education in Canada: A policy analysis." In T. Nesbit, S. M. Brigham, N. Taber, & T. Gibb (Eds.), *Building on critical traditions: Adult education and learning in Canada* (pp. 258–269). Toronto: Thompson Educational Publishing.

Goodall, J. (2005). *Harvest for hope: A guide to mindful eating.* New York, NY: Warner Books.

Goodman, M. K. (2004). Reading fair trade: Political ecological imaginary and the moral economy of fair trade foods. *Political Geography, 23*(7), 891–915.

Gussow, J. D. (2001). *This organic life: Confessions of an urban homesteader.* White River Junction, VT: Chelsea Green Publishing.

Guthman, J. (2008). Neoliberalism and the making of food politics in California. *Geoforum, 39*(3), 1171–1183.

Habermas, J. (1978). *Knowledge and human interests.* Boston, MA: Beacon Press.

Hall, B. (2006). Social movement learning: Theorizing a Canadian tradition. In T. J. Fenwick, T. Nesbit, & B. Spencer (Eds.), *Contexts of adult education: Canadian perspectives* (pp. 230–238). Toronto: Thompson Educational Publishing.

Hall, B., & Clover, D. (2005). Social movement learning. In L. M. English (Ed.), *International encyclopedia of adult education* (pp. 584–589). New York, NY: Palgrave Macmillan.

Hamilton, C. (2001). The case for fair trade. *Journal of Australian Political Economy, 48* (December), 60–72.

Harvey, D. (2006). Neo-liberalism as creative destruction. *Geografiska Annaler, 88B*(2), 145–158.

Harvey, D. (2005). *A brief history of neoliberalism.* Toronto: Oxford University Press.

Hauter, W. (2012). *Foodopoly: The battle over the future of food and farming in America.* New York, NY: The New Press.

Holt-Giménez, E., & Shattuck, A. (2011). Food crises, food regimes and food movements: Rumblings of reform or tides of transformation? *Journal of Peasant Studies, 38*(1), 109–144.

Holt-Giménez, E. (2009). From food crisis to food sovereignty: The challenge of social movements. *Monthly Review, 61*(3), 142–156.

International Assessment of Agricultural Knowledge, Science and Technology for Development (IAASTD). (2009). *Agriculture at a crossroads.* International Assessment of Agricultural Knowledge, Science and Technology for Development, Volume IV: North America and Europe. Washington, DC: Island Press.

International Council for Science. (2002). *Science and traditional knowledge.* Report from the ICSU Study Group on Science and Traditional Knowledge. Retrieved from http://www.icsu.org/publications/reports-and-reviews/science-traditional-knowledge/Science-traditional-knowledge.pdf

Jaffee, D., Kloppenburg, J. R. Jr., & Monroy, M. B. (2004). Bringing the "moral charge" home: Fair trade within the North and within the South. *Rural Sociology, 69*(2), 169–196.

Kingsolver, B. (2007). *Animal, vegetable, miracle: A year of food life.* New York, NY: HarperCollins.

Kloppenburg, J. Jr., Hendrickson, J., & Stevenson, G. W. (1996). Coming in to the foodshed. *Agriculture and Human Values, 13*(3), 33–42.

Koç, M., Sumner, J., & Winson, A. (Eds.). (2012). *Critical perspectives in food studies.* Toronto: Oxford University Press.

Maturana, H. R., & Varela, F. J. (1987). *The tree of knowledge: The biological roots of human understanding.* Boston, MA: Shambala Press.

McBride, M. F., & Burgman, M. A. (2012). What is expert knowledge, how is such knowledge gathered, and how do we use it to address questions in landscape ecology? In A. H. Perera, C. A. Drew, & C. J. Johnson (Eds.), *Expert knowledge and its application in landscape ecology* (pp. 11–38). New York, NY: Springer.

McMichael, P. (2009). A food regime analysis of the "world food crisis." *Agriculture and Human Values, 26*(4), 281–295.

Metcalf Foundation. (2008). *Food connects us all: Sustainable local food in southern Ontario.* Retrieved from http://metcalffoundation.com/publications-resources/view/food-connects-us-all-sustainable-local-food-in-southern-ontario/

Miles, A. (1996). Adult education for global social change: Feminism and women's movement. In P. Wangoola & F. Youngman (Eds.), *Towards a transformative political economy of adult education: Theoretical and practical challenges* (pp. 277–292). DeKalb, IL: LEPS Press.

Morris, R. K. (2005). Social movements. In L. M. English (Ed.), *International encyclopedia of adult education* (pp. 589–594). New York, NY: Palgrave Macmillan.

Morrison, D. (2011). Indigenous food sovereignty: A model for social learning. In H. Wittman, A. A. Desmarais, & N. Wiebe (Eds.), *Food sovereignty in Canada: Creating just and sustainable food systems* (pp. 97–113). Halifax: Fernwood Publishing.

Morrow, R. A., & Torres, C. A. (1995). *Social theory and education: A critique of theories of social and cultural reproduction*. Albany: State University of New York Press.

Norman, R. (2013). The fair trade movement. In J. P. Clark & C. Ritson (Eds.), *Practical ethics for food professionals: Ethics in research, education and the workplace* (pp. 203–220). Hoboken, NJ: Wiley-Blackwell.

Oxford University Press (OUP). (2007). Oxford word of the year: Locavore. Oxford University Press. Retrieved from https://blog.oup.com/2007/11/locavore/

Patel, R. (2007). *Stuffed and starved: Markets, power and the hidden battle for the world's food system*. Toronto: HarperCollins.

Pollan, M. (2006). No bar code: The next revolution in food is just around the corner. *Mother Jones*, May/June, 36–45.

Pollan, M. (2010, June 10). The food movement, rising. *New York Review of Books*. Retrieved from http://michaelpollan.com/articles-archive/the-food-movement-rising/

Ransom, D. (2001). *The no-nonsense guide to fair trade*. Toronto: Between the Lines.

Raynolds, L. T. (2002). Consumer/producer links in fair trade coffee networks. *Sociologia Ruralis*, *42*(4), 404–424.

Roberts, W. (2013). *The no-nonsense guide to world food*. Toronto: Between the Lines.

Russell, B. (1967). *Power: A new social analysis*. London: Unwin Books.

Sachs, W. (Ed.). (2007). *The development dictionary: A guide to knowledge as power*. London: Zed Books.

Smith, A., & MacKinnon, J. B. (2007). *The 100-mile diet: A year of local eating*. Toronto: Random House.

Sole, D., & Edmondson, A. (2002). Situated knowledge and learning in dispersed teams. *British Journal of Management*, *13*(S2), S17–S34.

Solomon, N. (2005). Knowledge. In L. M. English (Ed.), *International encyclopedia of adult education* (pp. 335–338). New York, NY: Palgrave Macmillan.

Stone, D. (2000). Knowledge, power and policy. In D. Stone (Ed.), *Banking on knowledge: The genesis of the global development network* (pp. 241–258). New York, NY: Routledge.

Sumner, J. (2006). From the knowledge economy to the knowledge commons: Resisting the commodification of knowledge. In G. Laxer & D. Soron (Eds.), *Not for sale: Decommodifying public life*, (pp. 203–217). Toronto: Broadview Press.

Sumner, J. (2008). Eating as a pedagogical act: Food as a catalyst for adult education for sustainability. *Kursiv—Journal fuer politische Bildung*, *4*, 23–37.

Walters, S. (2005). Social movements, class, and adult education. In S. B. Meriam & A. P. Grace (Eds.), *The Jossey-Bass reader on contemporary issues in adult education*, (pp. 138–148). San Francisco, CA: Jossey-Bass.

Winson, A. (2010). The demand for healthy eating: Supporting a transformative food "movement." *Rural Sociology 75*(4), 584–600.

Wittman, H., Desmarais, A. A., & Wiebe, N. (Eds.). (2010). *Food sovereignty: Reconnecting food, nature and community*. Black Point, Nova Scotia: Fernwood Publishing.

Chapter 5

Modern Southern Food: An Examination of the Intersection of Place, Race, Class, and Gender in the Quest for Authenticity

Kaitland M. Byrd

Introduction

Southern food has become popular in cities across the United States. Restaurants from coast to coast are advertised as providing "authentic" and "modern" Southern food. The question then becomes: for whom is this food authentic and modern? Authenticity in food is often accompanied by some degree of cultural appropriation. As such, what is authentic reflects changing social boundaries (Johnston & Baumann, 2010). Southern food, like the South in general, has a complicated history of oppression and exploitation. The examination of Southern food offers a lens for exploring the cultural boundaries between social groups. Studying food makes it possible "to learn what has gone on in the kitchen and the dining room—and what still goes on there . . . to discover much about a society's physical health, its economic condition, its race relations, its class structure, and the status of its women" (Egerton, 1987, p. 3). Therefore, food is an integral tool for understanding social processes.

This chapter explores race, class, and gender influences on the relationship between authenticity and appropriation in the creation of Southern foodways—as they are known today. This chapter first examines the relationship between the creation and consumption of Southern food, both historically and contemporarily. Focusing on the intersection of food, space, and inequality, the chapter then explores the culture of authentic Southern food, while acknowledging the role of appropriation. By analyzing barbecue as a symbol of Southern foodways, along with cookbooks as a representation of authenticity, the intersection of inequality and culture will emerge. Examining barbecue and cookbooks provides examples of how social and cultural boundaries intersect but remain dynamic. This chapter concludes by discussing the constructed boundaries of authenticity, as they exist in modern Southern foodways.

In the US South, social boundaries continue to be fixed along racial lines; and though structural barriers continue to exist across those boundaries, food remains a vehicle for interaction. African American influence on Southern food began with the slave trade as Africans were forced to adapt to the food products available in the United States while maintaining traditional cooking techniques. What emerged was a blend of African cooking practices with southern ingredients, some of which had come from Africa; this medley became the foundation of Southern foodways (Harris, 2011). Thus, what constitutes authentic Southern food is rooted in the history of racial oppression and the appropriation of foodways that began with African American women working in the kitchens of white plantation owners. Appropriation occurs when a dominant group, in this case white southerners, claims ownership over the cultural products (e.g., Southern food and the associated recipes) of another group, in this case African Americans. After slavery, African American women sought work as domestics in urban areas both inside and outside the South. As a result, what is considered authentic Southern food is a product of the interaction between whites and African Americans in the plantation environment and in kitchens (Ferris, 2013). Thus, the present conceptualization of modern Southern food is rooted in a history of appropriation, which attempts to create a shared history by glossing over the reality of who has historically created and prepared Southern food. Though Southern food was developed within cross-racial interactions, there is still a sharp contrast in the popular media and in popular opinion regarding who can cook authentic Southern food and who has traditionally cooked Southern food. Contemporary chefs choose to identify their food and restaurants in ways that convey ties to traditional southern foodways, while allowing themselves some artistic leeway in what is produced—as long as it still can be identified as "southern."

Previous research examines the intersection of these dynamics through historical cookbooks. Yet cookbooks can selectively highlight or omit aspects of Southern food. For example, depending on the audience, cookbooks have highlighted women as the source of recipes or have focused on professional male chefs as the source of cooking knowledge. Both versions tend to omit the influence of African American cooks (Fertel, 2013; Sharpless, 2013). Although the image of an African American woman as a cook has become iconic in the South through stereotypical images of African American women as mammies (Collins, 2008), African American women's influence on creating Southern foodways is often overlooked (Fertel, 2013).

Like cookbooks, barbecue offers an example of how history and culture intersect to create place-specific distinctions in food culture. Despite the present-day conceptualizations of barbecue—such as in barbecue competitions—the dish's past is rooted in racial oppression. Historically, slaves prepared barbecue for large gatherings held on plantations (Opie, 2008). Southern barbecue has influenced the creation of identities and culture in the South. Different barbecue styles are unique to various locations, and an affinity for a certain barbeque style creates a sense of location-based identity. Barbecue techniques are passed down through generations that occupy the same social and cultural space (Veteto & Maclin, 2011). In this way barbecue provides a case study of how important geographic specificity is to Southern food and the importance of place in understanding these differences.

What Is the South?

To understand Southern food, what constitutes the South needs clarification. Historically, there is disagreement over which states are included in the South. One definition of the South is any state that was in the Confederate States of America. Alternatively, the South can be expanded to include Oklahoma and Texas as well as Delaware, the District of Columbia, and Maryland. Or it may not include any state north of Virginia (Reed, 1986).

Reed (1986) argues that Southern culture is rooted in the particularism dominating the South. Americans do not often identify themselves in relation to a particular state or location (Reed, 1986). However, the opposite is true in the South. Local and regional location shape Southern identity and culture through the strong emotional connection that people who live—or have lived—in the South, have with a particular place. There is a degree of possessiveness that people in the South have regarding the place they live in. This possessiveness, or localism, is not commonly found in other parts of the country (Reed, 1986). This pattern of Southern localism continues to influence southern culture in terms of religion, violence, and stereotypes (Reed, 1986).

A more recent view argues that the South exists wherever southern culture is present (Ferris, 2013). Ferris's (2013) definition will be the one used in this chapter. Neither the South nor southern culture is confined to a specific physical location, thus place and culture must be conceptualized to recognize their spread beyond traditional bounded physical locations. Southern food has a strong history, and consumption is frequently limited by geography. Latshaw (2013) found that living in the South or being raised in the South are strong predictors of Southern food consumption. Although, as Ferris (2013) argues, the South, as a geographically bounded location, is inadequate to explain the consumption of southern culture generally, and Southern food specifically.

The concept of place is rooted in a shared understanding of the past. As change occurs more rapidly and more emphasis is placed on the future, the past becomes an anchor (Anderson, 2001). The past that is necessary for a sense of place is a product of collective memory where "events are selected, reconstructed, maintained, modified, and endowed with political meaning" (Said, 2000, p. 185). The reconstruction of the past becomes necessary to maintain a shared cultural identity in the midst of such social change. Cultural identity accounts for shared meanings that are a product of groups of people experiencing the same social conditions and structures (Alexander, 2007). Additionally, the past gives a degree of credibility to a place through the stories people living there tell about the place. Stories allow people to understand the place while also establishing a cultural identity associated with that particular place (Bird, 2002). People can influence the cultural identity of a place, and when different groups contest space, the identity of that place will be different for both groups. With this in mind, the South has historically been a contested space because of racial oppression and continuing inequality.

Southern Foodways

Race, class, and gender—in addition to place and space—continue to influence the creation and consumption of Southern food, creating distinctions in Southern foodways. Living in the South for an extended period of time, especially during childhood, increases one's consumption of Southern food, allowing it to become part of one's identity (Latshaw, 2013). Moreover, the relationship between people and food can influence their identity and their identification with aspects of southern culture. Although Southern food has a history associated with oppression and exploitation, it also offers the opportunity to create a shared sense of southern identity and community that is based on more equal power relations than those observed in the past (Egerton, 1987). One way this could happen is through a large amount of agreement across race, class, and gender about the importance of Southern food for a southern identity. The quest for authenticity in Southern food is rooted in the South's complex history of racial oppression. Interactions across racial lines have created Southern foodways. By exploring race, class, and gender in cookbooks and barbecue, the relationship between identity, appropriation, and the drive for authenticity will become apparent.

Southern Food

Now that our definition of the South has been clarified, I will define Southern food specifically. First, there is debate about the terms *southern cooking* and *soul food*. Miller (2013) distinguishes between the two terms, based on conceptions of racial superiority. Before the Civil War, white southerners drew distinct boundaries between what whites ate and what slaves ate. One phrase used to describe this was that whites "ate high on the hog," meaning that whites ate select cuts of meats that were often influenced by traditional European dishes. In reality, these distinctions were mostly illusions, particularly when white houses relied on slave labor to cook every meal. After the Civil War, this illusion was shattered because of the scarcity of food in the South. Typically, soul food is associated with several themes: ingenuity, community, resourcefulness, racial stigma, the social position of African Americans, and pork as a central dish. These conceptualizations of soul food, like other symbolic boundaries, do not consistently hold up under closer scrutiny (Miller, 2013). In general, soul food accounts for the racial dynamic of food, with African Americans having the most influence on Southern foodways (McCann, 2009). Despite the misconceptions and misinformation about the role of African Americans in Southern food (Hess, 2002), the link between race and food in the South cannot be ignored. Edna Lewis, one of the most influential African American female chefs, states, "The early cooking of Southern food was primarily done by African-Americans, men and women. It was then, and it still is now" (as cited by Ferris, 2013, p. 278).

Since racial boundaries are blurred and the distinction between soul food and southern cooking is problematic, I will use the term "Southern food" to describe food linked to the US South. Egerton (1987) labels southern cuisine as "the most positive element of our collective character, an inspiring symbol of reconciliation, healing, and union" because it has the ability to "unlock the rusty gates of race and class, age and sex" (Egerton, 1987, pp. 2–3). A few examples of what constitutes Southern food include pork, grits, watermelon, black-eyed peas, chitterlings, fried chicken, barbecue, and okra (Edge, 2004; Harris, 2011).

Southern food is not solely the product of cross-racial interactions, because not all white southerners owned slaves. Class played a major role in what people ate historically, as it still does. Native Americans, poor whites, and Appalachian residents all influenced what is seen as modern Southern food. However, for the purpose of this chapter, the focus is being placed on the plantation South and the relationship between African Americans and white southerners in shaping Southern foodways.

Southern food offers a unique understanding of the interplay between white privilege and public and private space. Slaves, especially African American women, were placed in a unique position because of their role as domestic workers and cooks in white households (Sharpless, 2010). The delineations often placed on culture to divide groups (Fuchs & Marshall, 1998) were used during slavery to define what foods whites should consume and what foods African Americans should consume (Harris, 2011). Despite these outward conceptualizations of Southern food, the day-to-day reality was far different, featuring extensive overlap in the foods both groups consumed (Harris, 2011). Southern foods, once associated with slaves, have become the hallmarks of modern southern cuisine.

Authenticity

The underlying appropriation of African American food culture calls into question what is labeled "authentic" Southern food. Authenticity facilitates the creation of a sense of solidarity and conformity to distinguish social groups (Hughes, 2000). Authenticity cements the claim to ownership of cultural capital; it is a superculture that distinguishes groups that are based on what is ours and what is theirs. Identity then ties people to these groups. Authenticity labels culture as belonging to a specific group, making it possible for a group to possess it (Bourdieu, 1984; Johnston & Baumann, 2007; Johnston & Baumann, 2010). Through processes of authenticity, aspects of other cultures are borrowed and claimed as part of one culture (Harrison, 2008). Appropriation plays a dominant role in Southern food as a way to erase the African American cooks while taking into account the influence they had on Southern food. The resulting culture became legitimate for white Southerners and acts as a status marker.

One form of authentic Southern food is popularized by elite chefs such as Paula Deen and Emeril Lagasse. These chefs have taken traditional Southern ingredients and techniques and repackaged them to cater to a higher social class. Johnston and Baumann (2010) distinguish between types of authentic culture, which are rooted in the dominant group's ability to define what is authentic culture. Thus, these chefs, because of their prestigious social positions, are defining authentic southern food and selling it to the larger, non-Southern population.

Though it is important to define "authentic culture," it is also possible to fabricate that culture. Hughes's (2000) exploration of how country-music culture is a process of fabrication to create an identity can be compared to Southern food culture. Traditional Southern dishes have been appropriated and claimed by white Southerners despite its origins in plantation kitchens. One example of this fabrication and appropriation is in cookbooks published by white women in Southern communities. Sharpless (2013) examines a church community cookbook created by white women. This book included numerous recipes that had likely been appropriated from African American cooks, but it only credited white women for the recipes. Occasionally the book listed only a first name, suggesting this recipe might have come from an African American woman (Sharpless, 2013). The cookbook was an opportunity for white women to define who they were through the dishes they prepared. These self-definitions make it possible to see what counted as authentic culture for this group; and despite their influence, African American women were excluded from possessing part of that culture (Ghaziani, 2009). The status that authenticity provides a group—in this case, white and female cookbook authors—depends on who has access to the authenticity (Alexander 2007).

The Role of Appropriation in Southern Food

The influence of African Americans on food culture began with the transatlantic slave trade. Africans, as well as the produce they used for food, traveled across the Atlantic and found a home in the Americas. Amid the horrors of slavery, African Americans played a major role in the development of Southern foodways. Throughout the South, slaves worked in plantation kitchens; and depending upon the plantation, some were growing and cooking their own food. As a result of this intermixing, traditional African dishes made their way onto the tables of white Southerners (Harris, 2011). Although conditions in America made it impossible to grow a majority of traditional African produce, there were some—including okra, watermelon, and black-eyed peas—that grew well in the South and continue to be identified with the South and with African Americans. For example, okra acts as a thickener in traditional African dishes and is used for the same purpose in traditional Louisiana gumbo. Slaves had no access to some traditional African produce, so they found substitutes in ingredients prevalent in the South. For example, yams were replaced by sweet potatoes (Harris, 2011).

Numerous dishes made their way from Africa to white tables this way, via the slaves' cooking. Despite this direct influence African and African American foodways had on the South, the paths are often ignored because "the culinary omnivores that were the South Carolina plantocracy came to claim African inspired dishes like Hoppin' Jon, red rice, and roux-less Charleston gumbo as their own" (Harris, 2011, p. 72).

Both during and after slavery, women traditionally found their place in the kitchen as part of their domestic responsibility (Davies, 1996). African American women, as cooks, became iconic in the South. The image of the mammy negated the influence that African American women had contributed to the development of Southern food. The role of African American women in creating traditional Southern foodways is often overlooked despite this profound influence (Sharpless, 2010).

Though slavery ended, and though white women assumed much of the responsibility for their own homes and families, many still wished to hire an African American cook after the war. Migration to urban areas and new freedom forced many African American women into employment. Numerous emancipated women found employment as domestic workers, especially as cooks. Cooking was understood as a low-status position because of its links to basic necessity and raw materials. White women tended to hire African American women when possible so that they did not have to fill these roles themselves (Sharpless, 2010).

The African American influence on food in the United States was not limited to the kitchens of the south. Southern food became a melding of products, techniques, and styles reflecting the relocation of Africans to the Caribbean, the US South, and eventually to the North. The spread of Southern food around the country followed the Great Migration, when large numbers of African Americans left the South after the Civil War and later (McCann, 2009; Miller, 2013;

Sharpless, 2010). In northern cities such as Philadelphia, free African Americans sold food on the streets and in some cases owned restaurants. Street vendors sold food adapted from traditional African cuisine, with the added spices typical of Caribbean cuisine, using products found in the United States (Harris, 2011).

As black women worked in kitchens to prepare their version of Southern food, cookbooks that focused on Southern food began to appear. Cookbooks offer the chance to explore what is deemed authentic Southern food because of the labels and locations given to them. With authenticity comes appropriation. This is not to suggest that chefs are not responsible for the dishes they create, but to show how "Southern food" has a more convoluted past than is generally acknowledged. Cookbooks offer a place to study the role of appropriation in Southern food; and since white women had access to resources, they were the ones who published the early cookbooks. These white women were typically not working in the kitchen but instead were relying on hired, often African American women to cook for them (Sharpless, 2013).

Two examples of appropriation in southern cookbooks can be seen in Fertel's (2013) exploration of the history of *The Picayune's Creole Cook Book* from Louisiana, and Sharpless's (2013) look at the *St. Paul's Episcopal Church Cookbook* from Texas. The recipes in various editions of the St Paul's cookbook are rarely credited to African American women, and when they are, appear only with a first name, whereas those linked to white women appear with first and last names. Aside from the recipes, the line drawings in the church cookbook and cover art in *Picayune's* conjure images of the "Old South" with African American servants. In the editions of *Picayune's* that were published in the 1920s, the front cover portrays an African American mammy. These editions also give more agency to African American cooks, though the agency is limited by the emphasis placed on the mammy image. In the 1940s edition, the cover art is hidden by a dust jacket. The dust jacket features a Eurocentric male chef, and the emphasis previous editions had placed on African American influence is replaced by a focus on male chefs (Fertel, 2013). Even when African Americans are given credit for their influence on Southern food, recipes, and techniques, it is often negated by slave imagery, especially mammy imagery. This negation makes it possible to appropriate the dishes African American cooks developed in the South during the eighteenth and nineteenth centuries. Suddenly, the African American chefs seem less important and have lower status than do white chefs (Fertel, 2013).

Cookbooks demonstrate how authenticity is mediated through appropriation. Another media example is modern chefs' cooking personae, which—under examination—demonstrate how culture develops through constant managing on the part of a dominant group. The consequences of this manipulation can be seen on the Food Network as well as in numerous cookbooks that have been printed both historically and more recently. Collins (2010) argues that the role of the media in portraying food and cooking began to take shape in the 1950s with the rising popularity of Julia Child. The concept of distinction in food consumption began to appear as professional chefs gained national and international fame and opened restaurants that catered to the elite. The birth of the Food Network created television stars who were also professional chefs. However, there were sharp racial and gender divides in what the network portrayed. Women, including Rachael Ray and Paula Deen, identified themselves as cooks instead of chefs. There is also an underrepresentation of chefs who are not white males. Although race is often overlooked on the Food Network, race and ethnicity play an important role, as chefs are often steered into cooking the type of food that matches their appearance and are met with resistance when they deviate from this standard (Inness, 2006). Again, distinction in "authentic" food culture is not only in the ingredients, but also in who prepares the food.

The media, particularly the Food Network, creates an elite food culture that is based on the stories chefs tell regarding their background. These stories highlight the uniqueness of the individual and often focus on the individual's biography as a justification for the authenticity of his or her food (Collins, 2009). For example, self-taught chefs are held in high esteem because their talent is often accompanied by a rising-out-of-poverty story. People can relate to or aspire to these narratives as the main character rises, despite the collective antagonism of society or poverty (Ewick & Sielby, 2003). These stories also convey authenticity to the position of the chef as an artist. The role of the biography conforms to Fine's (2003) conceptualization of authenticity. It is because of the artist's biography that the art is authentic. Fine (2003) argues that it is when the artist becomes more concerned with making a profit than with the art itself that authenticity disappears. Authentic art requires the artist to ignore the profits gained from the artwork. Therefore, self-taught chefs arguably become the symbol of authentic Southern food because they have a specific biography to support their claim of authenticity.

These biographies are also a mechanism that reinforces inequality. For example, the typical story involves a young boy learning how to cook from his grandmother, then working his way up to a chef position. These biographies regulate women to the background though they remain the bearers of tradition. The media emphasizes male chefs as the ones who can achieve elite status. Women who achieve elite status are labeled cooks, while men are labeled chefs. Media outlets such as the Food Network also steer chefs toward cooking the type of food that matches their perceived race or ethnicity and the sound of their last name (Collins, 2009).

The Food Network is an example of food that operates within the cultural realm of authenticity and appropriation. Restaurants also attempt to use authenticity and preexisting formulas as means to determine the future success of a restaurant. This authenticity is then created through labels and dishes that are deemed authentic and part of a preexisting tradition. Since authenticity in a specific food culture is ambiguous, specific dishes become the method of operationalizing a specific cuisine. The dishes, despite their true origins, represent the embodiment of Southern food; and people's responses to those dishes can be observed on the basis of frequency of consumption (Mohr, 1998).

Although the Food Network has been influential in shaping the perceptions of food culture throughout the United States, chefs of small, local, and large elite restaurants have also been influential in shaping what is considered Southern food. Restaurants across the South choose to remain traditional in terms of ingredients and final products while others seek authenticity through modern twists on traditional dishes. This idea of the modern-traditional restaurant balances regional dishes with other multiethnic influences. For example, at Empire State South in Atlanta, Georgia, a Canadian chef from San Francisco cooks Southern dishes with a European flair (Addison, 2013). Such mash-ups are occurring more frequently in the South and in urban areas around the country, while other restaurants are maintaining the same food, style, and menus they have employed for decades (Trillian, 2013). Despite the seemingly drastic differences between the traditional and the modern, these two types of restaurants showcase how social boundaries are flexible and constantly changing (Anderson, 2013; Reed & Reed, 2008; Sauceman, 2010). The dynamic changes are best illustrated through a discussion of barbecue's complex history.

Barbecue: The Intersection of History and Culture

Like all other aspects of Southern food, barbecue is continually changing and reinventing itself while remaining true to what historically made it distinct (see Elie, 2004; Veteto & Maclin, 2011). Barbecue has a past that is rooted in racial oppression and interactions that cannot be ignored when trying to understand the present version. Barbecue offers a case study of how history and culture intersect on a daily basis and constantly change the social boundaries that created it.

Barbecue is an example of the African slaves' influence on southern cuisine and how white southerners appropriated the slaves' foodways. Although there is some debate over the origins of the techniques and terminology of barbecue, researchers have traced these back to Native Americans and Taino Indians in the Caribbean. One contested explanation emphasizes European influence while ignoring other possibilities. It argues that barbecue in the United States began in the Virginia colony, with its origins in the English culture of boiling and roasting meats. Pigs were also widely available in the South, and as settlers traveled across the South, they brought barbecue with them (Moss, 2011). Despite barbecue's debated origins, it is still a symbol of Southern food culture. The methods of cooking have changed to reflect technological inventions that speed up cooking time, but the traditional method that takes days to complete is still revered. Though barbecue is becoming a popular pastime throughout the South, there is still the assumption that African American men make the best barbecue (Edge, 2011).

Barbecue was a popular social aspect of plantation life in the South. Whites held barbecues for huge social gatherings, and slaves, typically male, were responsible for preparing and cooking the meat. Slaves were often supervised by "an older slave who was recognized as the local barbecue master" (Moss, 2011, p. 29). This preparation involved an all-day-and-night process of roasting the meat over hot coals. Barbecues were used as a method of rewarding slaves on holidays and for political gatherings. Slaves were responsible for preparing barbecue, but despite this influence upon what is considered great Southern food, there are very few elite African American chefs. Instead, African American chefs are often relegated to less visible roles within elite kitchens. On the local level, they are often the hallmark of tradition and are sought out for their knowledge of techniques that can be traced back for hundreds of years (Edge, 2011).

Southern barbecue has been essential to the creation of identity and culture in the South. The dish has a strong reliance on place, with various cooking styles, sauces, and parts being used, depending on the location. The sauces as well as the techniques descend from African styles of cooking, as the barbecue cooks "[illegible] techniques passed down from West African ancestors, such as using lemon juice and hot peppers as essential ingredients. As in Africa, sauce recipes differed across the South" (Opie, 2008, p. 107). Different styles of barbecue are unique to different locations, and an affinity for a certain barbecue style creates a sense of identity that is based on that particular location (Veteto & Maclin, 2011). Barbecue techniques become linked to a particular place and are passed down through generations of people who continue to occupy that same space (Edge, 2011; Opie, 2008). Smith (2004) explains how the differences in sauces, wood, and types of meat depended upon the price and availability of a product in a specific location. For example, Texas uses beef, Kentucky uses mutton, and the rest of the South uses pork. The wood used in the smokers also varied, based on the availability of oak, hickory, sassafras, or mesquite. The main ingredient in the sauce was also varied to include pepper, vinegar, tomato, or mustard, with each region having a specific preference (Elie, 2004; Smith, 2004).

Despite the importance of barbecue in Southern foodways, the slaves' role in developing barbecue is often ignored (Warnes, 2008). Like other aspects of Southern culture, barbecue is a product of the interactions between racial groups over time, and even though whites continue to have problems admitting this influence, it still exists (Ferris, 2013; Walsh, 2004). Confederate flags have even been known to fly at barbecues to convey that African Americans do not belong at such events (Walsh, 2004).

Barbecue is one food that has all the hallmarks of Southern food. It began with the slave trade, traveled to the United States on slave ships, and was prepared by slaves for white plantation owners. The cut of meat, style of sauce, and type of wood all identify barbecue with a specific geographical region, even though today those techniques have spread to other locations. The quest for authentic barbecue is based on preconceived ideas about style and regionalism. Barbecue, in various forms, now appears on the Food Network and in the cookbooks and restaurants of elite chefs. Barbecue, like other Southern foods, continues to change while remaining true to its history of place, inequality, and authenticity.

Conclusions and Observations

Southern food, like the South as a whole, has been branded, commodified, and distinguished from other regions to the point that what is an "authentic" southern identity has become detached from the historical reality of the South (Cox, 2008). The study of Southern food as appropriation offers a new lens for exploring what constitutes a modern Southern culture and identity. Studying food makes it possible "to discover much about a society's physical health, its economic condition, its race relations, its class structure, and the status of its women" (Egerton, 1987, p. 3).

Southern food has been confined to a particular place, where all these categories can overlap, interact, and blur the social boundaries that modern rules try to establish. As a result of this lack of clear boundaries, authenticity has arisen as a way to determine who has ownership over Southern foodways. This quest for authenticity has followed the same path that current trends try to enforce, with the dominant group using their position to take credit for creations in the kitchen while other groups, such as African Americans, do not have the same ability.

This is an example of why stories are so important for creating authenticity. Despite the manipulation of stories attempting to erase the influence of African American cooks, their influence upon Southern food runs so deep within Southern foodways that it is not possible to completely erase it. This is why Southern food is seen as a way to overcome past transgressions, because stories can evolve and return credit to the African American cooks who shaped Southern food.

Regional foodways are not limited to the South. Throughout the world, regional foodways are linked to a people and a place. The direct link between people and place becomes more problematic in the modern, globalized world. Despite these changes, food in the modern world offers a valuable lens for looking at geographic locations, the people who call a particular location "home," and the inequality that continues to structure social relations around the world. In the modern global context, strict geographical boundaries become less prevalent than symbolic boundaries. Food offers a way to bridge the gap between geographic and symbolic boundaries. The quest for authentic food demands an awareness of the history of a place and of its people. This history provides the opportunity for a person to find authentic food, even when outside the traditional geographic boundary. For example, a person does not have to be in the South to consume Southern food; but to consume Southern culture, or food, an awareness of the place and that place's history are important dimensions that allow a person to decide what is authentic, instead of relying on a label.

Food is a powerful tool for understanding society. It is rooted in place-based struggles of and between authenticity, racial oppression, gender, historical tradition, and an ever-changing culture that is continually shaped by globalization. Understanding the foodways of a specific place provides researchers with a lens to study the social relationships and structure of a specific place. Food is also a necessary and integral part of life, making it possible to use place-based perspectives of food to compare various locations both historically and in the modern world.

References

Addison, B. (2013). Empire state south: Athens star chef Hugh Acheson brings Atlanta its latest southern sensation. In B. Anderson (Ed.), *Cornbread nation: The best of southern food writing* (Vol. 6, pp. 189–191). Athens: University of Georgia Press.

Alexander, J. C. (2007). The meaningful construction of inequality and the struggles against it: A 'strong program' approach to how social boundaries change. *Cultural Sociology, 1*(1), 23–30.

Anderson, B. R. (2001). Western nationalism and Eastern nationalism: Is there a difference that matters? *New Left Review, 9*, 31–42.

Anderson, B. (Ed.). (2013). *Cornbread nation: The best of southern food writing* (Vol. 6). Athens: University of Georgia Press.

Bird, S. E. (2002). It makes sense to us: Cultural identity in local legends of place. *Journal of Contemporary Ethnography, 31*(5), 519–547.

Bourdieu, P. (1984). *Distinction: A social critique of the judgment of taste.* (R. Nice, Trans.). Cambridge, MA: Harvard Press.

Brett, J. (2012). Conclusions: Culture, tradition, and the political economy. In E. Finnis (Ed.), *Reimagining marginalized foods: Global processes, local places* (pp. 156–166). Tucson: University of Arizona Press.

Collins, K. (2009). *Watching what we eat: The evolution of television cooking shows.* New York, NY: Bloomsbury.

Collins, P. H. (2008). *Black feminist thought.* New York, NY: Routledge.

Cox, K. L. (2008) Branding dixie: The selling of the American south. In A. J. Stanonis (Ed.), *Dixie emporium: Tourism, foodways, and consumer culture in the American South* (pp. 50–68). Athens: University of Georgia Press.

Davies, C. (1996). The sociology of professions and the profession of gender. *Sociology, 30*(4), 661–678.

Edge, J. T. (2004). *Fried chicken: An American story.* New York, NY: G. P. Putnam's Sons.

Edge, J. T. (2011). Patronage and the pits: A portrait, in black and white, of Jones Bar-B-Q Diner in Marianna, Arkansas. In J. R. Veteto & E. M. Maclin (Eds.), *The slaw and the slow cooked: Culture and barbecue in the Mid-South* (pp. 43–50). Nashville, TN: Vanderbilt University Press.

Egerton, J. (1987). *Southern food: At home, on the road, in history.* Chapel Hill: University of North Carolina Press.

Elie, L. E. (Ed.). (2004). *Cornbread nation: The United States of barbecue* (Vol. 2). Chapel Hill: University of North Carolina Press.

Ewick, P., & Silbey, S. (2003). Narrating social structure: Stories of resistance to legal authority. *American Journal of Sociology, 108*(6), 1328–1372.

Ferris, M. C. (2013). The "stuff" of southern food: Food and material culture in the American South. In J. T. Edge, E. Engelhardt, & T. Ownby (Eds.), *The larder: Food studies methods from the American south* (pp. 276–311). Athens: University of Georgia Press.

Fertel, R. T. (2013). "Everybody seemed willing to help": The Picayune Creole Cook Book as battleground, 1900–2008. In J. T. Edge, E. Engelhardt, & T. Ownby (Eds.), *The larder: Food studies methods from the American south* (pp. 10–31). Athens: University of Georgia Press.

Fine, G. A. (2003). Crafting authenticity: The validation of identity in self-taught art. *Theory and Society, 32*(2), 153–180.

Fuchs, S., & Marshall, D. A. (1998). Across the great (and small) divides. *Soziale Systeme, 4*(1), 5–31.

Ghaziani, A. (2009). An "amorphous mist"? The problem of measurement in the study of culture. *Theory and Society, 38*(6), 581–612.

Harris, J. B. (2011). *High on the hog: A culinary journey from Africa to America*. New York, NY: Bloomsbury.

Harrison, A. K. (2008). Racial authenticity in rap music and hip hop. *Sociology Compass, 2*(6), 1783–1800.

Hess, K. (2002). Okra in the African diaspora of our south. In J. Egerton (Ed.), *Cornbread nation: The best of southern food writing* (Vol. 1, pp. 240–252). Chapel Hill: University of North Carolina Press.

Hughes, M. (2000). Country music as impression management: A meditation on fabricating authenticity. *Poetics, 28*(2–3), 185–205.

Inness, S. A. (2006). *Secret ingredients: Race, gender, and class at the dinner table*. New York, NY: Palgrave MacMillan.

Johnston, J., & Baumann, S. (2007). Democracy versus distinction: A study of omnivorousness in gourmet food writing. *American Journal of Sociology, 113*(1), 165–204.

Johnston, J., & Baumann, B. (2010). *Foodies: Democracy and distinction in the gourmet foodscape*. New York, NY: Routledge.

Latshaw, B. A. (2013). The soul of the south: Race, food, and identity in the American South. In J. T. Edge, E. Engelhardt, & T. Ownby (Eds.), *The larder: Food studies methods from the American South* (pp. 99–127). Athens: University of Georgia Press.

McCann, J. C. (2009). *Stirring the pot: A history of African cuisine*. Athens: Ohio University Press.

Miller, A. (2013). *Soul food: The surprising story of an American cuisine, one plate at a time*. Chapel Hill: University of North Carolina Press.

Mohr, J. W. (1998). Measuring meaning structures. *Annual Review of Sociology, 24*, 345–370.

Moss, R. F. (2011). A history of barbecue in the Mid-South region. In J. R. Veteto & E. M. Maclin (Eds.), *The slaw and the slow cooked: Culture and barbecue in the Mid-South* (pp. 25–42). Nashville, TN: Vanderbilt University Press.

Opie, F. D. (2008). *Hog and hominy: Soul food from Africa to America*. New York, NY: Columbia University Press.

Reed, D. V., & Reed, J. S. (Eds.). (2008). *Cornbread nation: The best of southern food writing* (Vol. 4). Athens: University of Georgia Press.

Reed, J. S. (1986) *The enduring South: Subcultural persistence in mass society* (2nd ed.). Chapel Hill: University of North Carolina Press.

Said, E. W. (2000). Invention, memory, and place. *Critical Inquiry, 26*(2), 175–192.

Sauceman, F. W. (Ed). (2010). *Cornbread nation: The best of southern food writing* (Vol. 5). Athens: University of Georgia Press.

Sharpless, R. (2010). *Cooking in other women's kitchens: Domestic workers in the South, 1965–1960.* Chapel Hill: University of North Carolina Press.

Sharpless, R. (2013). The women of St. Paul's Episcopal Church were worried: Transforming domestic skills into saleable commodities in Texas. In J. T. Edge, E. Engelhardt, & T. Ownby (Eds.), *The larder: Food studies methods from the American south* (pp. 32–56). Athens: University of Georgia Press.

Smith, S. (2004). The rhetoric of barbecue: A southern rite and ritual. In L. E. Elie (Ed.), *Cornbread nation: The United States of barbecue* (Vol. 2, pp. 61–68). Chapel Hill: University of North Carolina Press.

Trillian, C. (2013). No daily specials. In B. Anderson (Ed.), *Cornbread nation: The best of southern food writing* (Vol. 6, pp. 198–205). Athens: University of Georgia Press.

Veteto, J. R., & Maclin, E. M. (Eds.). (2011). *The slaw and the slow cooked: Culture and barbecue in the Mid-South.* Nashville, TN: Vanderbilt University Press.

Walsh, R. (2004). Texas barbecue in black and white. In L. E. Elie (Ed.), *Cornbread nation: The United States of barbecue* (Vol. 2, pp. 48–60). Chapel Hill: University of North Carolina Press.

Warnes, A. (2008). *Savage barbecue: Race, culture, and the invention of America's first food.* Athens: University of Georgia Press.

Chapter 6

Distinction, Disdain, and Gentrification: Hipsters, Food People, and the Ethnic Other in Brooklyn, New York

Kathleen LeBesco and Peter Naccarato

Introduction

This chapter focuses on the recent culinary renaissance in the New York City (NYC) borough of Brooklyn, using the borough as a lens for exploring connections between food, place, and identity. It argues that specific Brooklyn neighborhoods have become contested spaces where different groups compete for status through their food practices. The first part of the chapter teases out the similarities and differences between hipsters and food people as each group vies for literal and imagined space within this changing Brooklyn landscape. Then, the chapter considers the impact of this influx of hipsters and food people on the ethnic, immigrant communities that have sustained many of these Brooklyn neighborhoods for generations. The overarching goal of the chapter is to better understand the social, political, and ideological power of food and foodways in a particular geographical and ideological space, Brooklyn.

Ultimately, our investigation of NYC hipster food discourse surfaces both the rhetorical function of the hipster figure in the food people imaginary and the role of the ethnic, immigrant "Other" for hipsters and food people alike. We contend that though very few people willingly identify as hipsters, the figure of the hipster exists both as a real and as an imagined identity. Given that one aspect of hipster identity is a commitment to individualism that shuns homogenizing group identifications, it is understandable that those who embrace stereotypically hipster lifestyles and values nonetheless refuse to self-identify, as such. In other words, anyone who stakes a claim to a hipster identity can be accused of undermining that very identity by publicly embracing it. Nonetheless, we assert that the hipster is a real identity category, the shared characteristics of which we discuss in more detail below. At the same time, we argue that the hipster also functions as an imagined figure that provides food people with a "straw man in skinny jeans" (Clayton, 2010, p. 30).

We use the category "food people" for individuals, typically of middle-class privilege, who either self-identify as or may be identified by others as "foodies." As we discuss below, we have found that the label "foodie" continues to carry positive connotations even as we are witnessing a turn against it by those who feel that it signifies a level of class and cultural pretension with which they do not want to be associated. Through food-culture fieldwork in Williamsburg, East Williamsburg, Downtown Brooklyn, and Cobble Hill, and through interviews with Brooklyn-based self-identified food people, we question, after Jace Clayton, the ways in which "criticizing the hipster is a way of discussing gentrification and neighborhood change—while exempting oneself from the process" (2010, p. 29). Because hipsters are more likely to reside in specific places, whereas food people are more likely to visit particular neighborhoods that are emerging as trendy culinary enclaves, we contend that hipsters serve as more immediate targets of criticism than food people in debates about the economic and racial implications of gentrification.

At the same time, these neighborhoods become spaces in which hipsters and food people meet, engaging in a complex dance of production and consumption. While one criticism often lobbed at hipsters is that they create identities solely through consumption, we question if such criticism is valid in relation to their food practices. As hipster neighborhoods attract not just restaurants but also small, independent food markets and other artisanal food businesses, they suggest a more nuanced relationship between production and consumption. We are particularly interested in the growth of artisanal and locally sourced food markets and restaurants in these hipster neighborhoods, arguing that they are indicators of how food people and hipsters both seek to acquire culinary capital, namely, the status that is acquired by participating in certain practices. But as they do so, they are not necessarily cognizant of how these food practices have long histories within ethnic, immigrant communities that are struggling to survive gentrification. Ironically, although hipsters and food people alike often trace their own food practices back to their real or imagined immigrant ancestors, they do so without necessarily recognizing the impact of the commodification of these ethnic foodways on today's ethnic, immigrant neighborhoods. What are we to make of the embrace of such foodways as signifiers of cultural distinction among food people and their imagined hipster counterparts, and can we locate in either a political program? For, as Ben Davis writes, without a political program, "Any critique simply falls into the trap best illustrated by the *Onion* headline: 'Two Hipsters Angrily Call Each Other Hipster'" (2013, p. 124).

Geographical Place, Ideological Space

We begin by thinking about Brooklyn's culinary and cultural history in the context of geography. As historian Anne Mendelson explains, after the retreat of the Wisconsin ice sheet around three million BCE, Greater New York was left with "the most stupendously plentiful, diverse, and annually sustained food supply of any area on the East Coast" (2009, p. 16). Such natural resources, when combined with New York City's role as a hub of industrialization and immigration throughout the nineteenth and early twentieth centuries, make possible a physical geography capable of producing a rich and diverse cultural and culinary landscape. At the same time, this geography introduces one of the central characteristics of Brooklyn's imagined identity, namely, its complicated—and often contentious—relationship to Manhattan. While separated geographically by the East River, the distance between these two boroughs encompasses more than just physical space. Described historically by Matt Levy—self-avowed Brooklyn hipster and partner in his family's business, Levy's Unique Tours—as a righteous, religious, pastoral "city of spires" that had a rivalry with money-oriented Manhattan (M. Levy, personal communication, March 19, 2014), Brooklyn continues to claim a higher moral ground in relation to Manhattan, which "existed to trade, to accumulate wealth, and to live well" (Grimes, 2009, p. 101). If much of the country in the nineteenth century viewed New York as an "amoral anomaly" in this otherwise "moralizing republic," Brooklynites extended this characterization further as they joined the rest of the country in regarding Manhattan "warily—although disapproval was mingled with desire" (Grimes, 2009, p. 101).

This history provides an important context for understanding the current ideological dynamics of Brooklyn's cultural and culinary reawakening. First, as certain Brooklyn neighborhoods emerge as hipster enclaves, they lay claim to a lifestyle and identity that has deep roots in New York City, particularly among those who sought "forms of rebellion against the materialistic values that America embodies"

(Parasecoli, 2009, p. 127). In seeking ways to "be more than their financial position allow[ed] them to be," hipsters embraced "a different kind of status symbol" that was "about being an insider, about being one step ahead of the rest of the flock" (Parasecoli, 2009, p. 127). However, as hip entered the mainstream, it "helped strengthen and sell materialistic and money-driven culture that it had originally tried to undermine" (Parasecoli, 2009, p. 127), ultimately becoming its own source of prestige to those who could afford its offerings. Of course, this included dining out as increasingly pricey hip restaurants became "accessible only to the few who [had] the economic means to pay for them and who [did] not always have the cultural capital necessary to enjoy them" (Parasecoli, 2009, p. 129). Such coopting of hip, coupled with the ever-rising cost of living in Manhattan, paved the way for Brooklyn's revitalization by a new generation of "authentic" hipsters in the borough that had always sought to define itself as different from—and superior to—Manhattan. However, because this claim to authenticity by Brooklyn hipsters depends on their embrace of its culinary renaissance, it also creates the contentious space where hipsters meet food people, a convergence that has important ramifications for many of Brooklyn's ethnic, immigrant neighborhoods.

To understand the relationship between physical geography and imagined space when it comes to this encounter between hipsters, food people, and their ethnic "Other," we borrow Joy Santloffer's notion of *asphalt terroir*. Traditionally used to define "the uniqueness of place, climate, and the producer's skills" that merge to "create distinctive tastes and products rooted to a locale" (2009, p. 175), Santloffer uses *terroir* to characterize the "ineffable bond" that links food producers to their New York City location. In some instances, this link aligns with a traditional sense of *terroir*, for example, when some New York food manufacturers point to the city's water as "an essential ingredient in their products" (2009, p. 192); but in other cases, such uniqueness is connected less to specific physical characteristics of New York geography and more to shared values, traditions, and aspirations, many of which serve to connect hipsters and food people insofar as they relate to the essential yet contentious claims to authenticity made by both groups.

It is this feature of *asphalt terroir* that Jennifer Causey uncovers in her interviews with several Brooklyn food artisans. Characterizing Brooklyn as a "diverse community filled with creative, motivated individuals, who are championing a return to craftsmanship and artisanal making" (2013, p. 11), Causey emphasizes the role of "place" in motivating the individuals who have created and are sustaining

Brooklyn's culinary renaissance. Such a characterization is extended beyond food artisans by Agatha Kulaga and Erin Patinkin, owners of Ovenly in Red Hook, who explain why Brooklyn is an essential ingredient in their success: "Brooklyn is home to a large group of young people who are trying to 'make it'—artists, musicians, chefs, techies, distillers, fashion designers, filmmakers, what have you. Everywhere you look, someone is working on an interesting project. It's inspiring" (Causey, 2013, p. 134). Thus, the physical place of Brooklyn serves as an incubator for likeminded individuals to bring their shared aspirations and values to bear on their medium of choice, from arts and crafts to food and beverages.

Hipsters and Food People

While the "culinary renaissance" in Brooklyn, heralded by the *New York Times* in 2009, has been multifaceted, one of its most notable attributes has been its relationship with the simultaneous embrace by many Brooklynites of hipster styles and identities. Like many American cities, New York has its own hipster enclaves: Williamsburg, a Brooklyn neighborhood, is currently considered NYC's main hipster hood. Mark Greif notes that the hipster moment "is strongly associated with neighborhoods in cities across the United States that represented either new zones of white recolonization of ethnic neighborhoods or subcolonies of established bohemian neighborhoods" (Greif, 2010a, p. 6). Often relocating postcollege, these hipsters are drawn to urban centers thick with culture where they can establish identities that reflect their shifting relationship to the mainstream. Having left the privileged space of the university, they find themselves somewhat disoriented as they seek to establish identities that assume a deeply ironic attitude toward conventional aspirations and values.

As Grief points out, the hipster "still possesses enormous reserves of what Pierre Bourdieu termed cultural capital, waiting to be activated—a degree, the training of the university for learning tiny distinctions and histories, for the discovery and navigation of cultural codes—but he or she has temporarily lost the real capital and background dominance belonging to his class. Certain kinds of subculture allow cultural capital to be re-mobilized among peers and then within the fabric of the 'poorer' city, to gain distinction and resist declassing" (Greif, 2010b, p. 160–161). We contend that hipsters who are temporarily denied status vis-à-vis access to economic capital seek to reassert their individual and social identities through the acquisition of culinary capital as they adopt "certain food practices [that] give [them] a sense of distinction within their communities" (Naccarato and LeBesco, 2012, p. 1).

Such connections between hipster identity and food practices are affirmed in *The Hipster Handbook*, which explains that hipsters do not dine at chain restaurants, unless kitsch is their aim (Lanham, 2002, p. 8). Hipsters are "always more culturally aware than most," enjoying "Tibetan, Vietnamese, Moroccan and American food with equal zest" and are savvy enough to know pinot noir from cabernet (Lanham, 2002, p. 13). They "prefer restaurants that write their menus on chalkboards, have low ceilings, use soft lighting, and have waiters with body odor" (Lanham, 2002, p. 29)—an aroma, that Lanham says, portends "authenticity."

Such disdain for fast-food chains claims cultural awareness, and the demand for culinary authenticity suggests strong parallels between the eating habits of hipsters and those embraced by self-proclaimed food people. However, while the category of "foodie" has attained a certain level of cultural cachet in recent decades, and some self-identified food people willingly describe themselves as foodies, the "hipster" is usually talked about with disdain; it is rarely a term of self-identification. Hipsters share foodies' critique of "the abysmal quality of mainstream food" and their interest in distinguishing themselves from others, based on knowledge, taste, and food practices (Johnston & Baumann, 2010, p. 53). Yet only hipsters, not foodies, are accused of "abandoning the claims of counterculture...while retaining the coolness of subculture" (Greif, Ross, & Tortorici, 2010, p. xvii). Yet as both hipsters and food people alike utilize their food practices in their self-fashioning, they do so without necessarily acknowledging the class privilege that allows them to do so:

> The new middle classes—tastemakers and trendspotters *par excellence*—make fullest use of their savvy in this setting, disdaining through the subtle movements of bodies and wallets the *passé* and *déclassé*. The game is always stacked in their favour; their lifestyle journalists know it, their restaurateurs know it, their 'purveyors of fine foods' know it...The hidden injuries of class are played out in the turf wars over taste, on home improvement shows and cookery pages, where symbolic violence comes on a plate or the table it sits on; the name we give it: *connoisseurship*. (Bell, 2002, p. 14)

As hipsters and food people both stake their claim to such connoisseurship, they reveal the social and ideological role that specific food practices can play in granting status to those who embrace them. At the same time, they reveal the extent to which this process plays out without acknowledging that other communities—often immigrant, ethnic communities—may have embraced similar food practices for generations but lack the cultural or class status to brand them as culinary trends.

To learn more about the hipster's ideological function in Brooklyn, we surveyed self-identified Brooklyn-based "food people" about identity and the borough's food scene. Ten respondents, recruited via friends who live or work in Brooklyn, filled out lengthy open-ended questionnaires about their impressions of foodies, hipsters, and foodways in the borough. They ranged in age from twenty-five to seventy-two; eight identified as white/Caucasian, one as biracial, and one as Latino. All were college graduates, with six having completed at least one graduate degree. Although identifying oneself as a "food person" was a requirement of participation in the survey, we were somewhat surprised about the ambivalence of our respondents to identify themselves explicitly as "foodies." About half had no problem with the term, but one of those who claimed to be identify as a foodie said, "I would never *refer* to myself as a foodie," hinting at a problematic connotation. Others rejected the term, stating: "I guess the ambivalence is that I think sometimes the whole scene is too precious and we are making too much out of a resource that is not equally affordable/accessible" and "to be a foodie suggests a level of passion that I don't feel." Of those who overtly addressed the connotations of the term "foodie," one thought "the word has a negative connotation and is associated with people who are supremely concerned with elements of the dining experience in which I'm not interested" while the other argued that "'Foodie' tends to have a neutral to positive valence (unlike 'hipster')."

In contrast, not a single one of our respondents identified as a hipster. Although one respondent attempted to resuscitate the figure of the hipster ("the people who hipster-bash are afraid they will read as hipster to someone else"), all but one other pointed to their disqualification on the basis of age ("I'm too old"). A few professed that they were not clear about what a hipster was, but those who seemed familiar with the term distanced themselves from it, noting, "I am actually rather reflective about my tastes, and modest, so there is no hope for me as a hipster," and "[I'm] too old, too poor, not nearly pretentious, cool, or 'stylish' enough. Do not have skinny jeans or a trust fund."

When asked plainly about the differences between foodies and hipsters, one respondent noted: "I think of hipsters as those invested in consumption, whereas I think of foodies as those invested in consumption as it is somehow tied to creation. This is not to say that all foodies are cooks, but I think foodies are interested in some way in how foods are made or paired or conceived of (there is an element of knowledge attached to the consumption that is essential to the enjoyment of that consumption)." This seems to be a common-enough notion, but it runs counter to our impressions from our hipster food fieldwork, which we discuss in the next section. Several suggested that the categories of the hipster and the foodie had some overlap in their shared area of interest (e.g., "both 'hipster' and 'foodie' are about making public consumption practices and making public aesthetic judgments"), but not in the attitude they brought to that interest.

When asked to describe the Brooklyn food scene, respondents praised its "artisanal/home-grown quality," its general quality and variety, and the bounty of ethnic cuisines and farmers markets available. The respondent who was most suspicious of "hipster-hating" pointed out that "the bougie places in Brooklyn are into authenticity (pretending globalization didn't happen, I guess)." Commenting on the relationship of the hipster food scene to the overall food scene there, he noted that gentrifiers were keen to patronize hip restaurants and various ethnic eateries: "I've never seen my 60-something neighbor from Grenada at our local farm-to-table bougie place, but I sometimes get jerk chicken at the place down the street that serves the Caribbean immigrants."

The one-way-ness of this relationship between "authentic" ethnic eats and what our respondent described as the "bougie" places frequented by foodies and hipsters alike resonates with Bell's summary of Ulf Hannerz (1990), who noted that "the endless safari of the cosmopolitan, searching out the exotic and the authentic, is essentially a predatory practice: the pillage of resources, the scouring of habitats, the uprooting and repackaging of the foreign, the novel, the dangerous" (Bell, 2002, p. 14). Of course, those with the requisite economic and cultural capital are the ones able to engage in such foraging. This may also help to explain the disdain for hipsters and their twee concerns that surfaced in our research with food people. Are these self-identified food people attempting to distance themselves from the racial, ethnic, and class implications of gentrification—the pricing out of the less wealthy in neighborhoods as property values increase—by offering up the hipster as their straw man? And if this is the case, can they be accused of throwing up the kind of smoke screen described by Anthony Galluzzo: "Better to focus on the kinds of things that suggest effete privilege—all [the hipster's] free time frivolously spent and with whose money?—than offer a critique of the truly privileged and the socioeconomic system which sustains their privilege" (2013, n.p.)? Of course, these food people would be implicated in such a critique, which would necessarily examine the structural factors surrounding cycles of disinvestment, gentrification, and development that have happened in Brooklyn and elsewhere on the backs of ethnic others.

Overall, the results of our survey reveal a familiar disdain for the "hipster," a somewhat ambivalent attitude toward the "foodie," and an acknowledgement that hipsters and foodies both share an ability to move seamlessly across the Brooklyn foodscape—from "bougie" hotspots to ethnic restaurants, from artisanal shops to immigrant enclaves. However, whereas they enjoy such culinary mobility, their ethnic, immigrant neighbors are less capable of making such forays into areas where gentrification has reconfigured class and culinary boundaries. As such, geographic place serves as a key component for understanding the ideological conflicts that are at play as the Brooklyn foodscape is significantly transformed. By studying these spaces, we are better able to tease out the racial, ethnic, and class implications of Brooklyn's much-touted culinary renaissance. These conclusions are affirmed by our research in some of these Brooklyn neighborhoods, which we discuss in the next section.

Touring Distinction: Hipster and Immigrant Food Scenes

For a firsthand experience of the hipster food scene in Brooklyn, we contacted Matt Levy of Levy's Unique Tours and requested a tour that highlighted the emerging food scene in Brooklyn. We found that both the tour he organized and his personal commentary throughout the day provided important insight into our investigation. With regard to the hipster figure, Levy both embodied it and offered critical insight into it as a self-consciously ironic identity. Levy met us wearing bright salmon-colored pants, a paint-smattered Miller Beer uniform work shirt with an embroidered "Javier" name tag, and a yellow plaid jacket. Sporting a handlebar moustache (to which he would make repeated reference throughout the day), he immediately commented on Peter's chain-store coffee cup: "Panera?! Why are you drinking Panera? There's going to be some *good* coffee on this tour!" We had found our food hipster.

Our day with Levy supported our conclusion that there is an important relationship between place, food, and identity in Brooklyn's culinary renaissance. Before delving into the current food scene, Levy reviewed Brooklyn's rich history, from its pivotal role in the Revolutionary War to its rich immigrant past. As he did so, his Brooklyn pride (sans any hipster irony) was clear, providing insight into one of the important connections between place and identity, particularly as it provided a context for framing Brooklyn's revitalized food scene. At the same time, the tour surfaced tensions vis-à-vis race, ethnicity, and class as components of this resurgence.

While Levy and several of the food purveyors we met over the course of the day sought to bridge the current Brooklyn foodscape with its past, such connections raise important questions. Shelsky's Smoked Fish, for example, begins its "story" as told on its website by identifying New York as a "city of immigrants" that has "many cuisines to which it can lay claim." Highlighting the influence of Eastern European Jewish immigrants to this diverse food scene, Shelsky's notes a severe decline of such "appetizing shops" in New York and then connects itself with the recent renaissance that has brought "some of New York's most exciting food" to Brooklyn. Although praising Brooklyn as a "foodie paradise," the website positions Shelsky's as responding to something that was missing, namely, "the Smoked Fish and Appetizing shop of old." Thus, Shelsky's positions itself with one foot in the past—bringing back the lost traditions of the Eastern European Jewish immigrant community—and one foot in the present and future as it plays a role in Brooklyn's recent cultural and culinary renaissance. In their words, Shelsky's is a "New-Brooklyn, updated nod to the traditional smoked fish and appetizing shops that dotted the Lower East Side and Brooklyn decades ago" (Shelsky's, 2014).

Implicit in Shelsky's story is a desire to construct a historical narrative that links food, ethnicity, and place as it revitalizes culinary traditions brought to particular neighborhoods in Brooklyn and Manhattan by Eastern European Jewish immigrants. In doing so, it establishes itself as a destination both for local and for long-distance customers who want to use their personal food practices to reinforce their own ethnic identities by connecting with their real or imagined immigrant ancestors. At the same time, Shelsky's is focused not merely on duplicating the past, but also on using it as a foundation for connecting with contemporary food trends. Thus, Peter took the opportunity to sample the whitefish sandwich, which begins with the traditional whitefish on a bagel, but updates it with cilantro and jalapeño peppers. While this is one indicator of Shelsky's desire to expand its customer base to include those food people who are seeking out trends that blend the traditional and the contemporary, there is another aspect of its product menu that reveals the class implications of this effort. With some prodding from Matt Levy, Peter Shelsky offered a taste of his best stuff, the Ruby Red Wild Caught smoked salmon, which he sells for over $85 per pound. While we could debate whether or not this delicacy tasted significantly better than the other smoked salmon options that filled the display case, we are more interested here in what the decision to carry such a product reveals about the tension between ethnicity, class, and the revitalized Brooklyn food scene. It is just as difficult to imagine the traditional smoked fish shops to which Shelsky's traces its roots carrying products with comparable price tags as it is to imagine first-generation Eastern European Jewish immigrants having the means to pay for such luxuries. And whether it is food people who romanticize their "authentic" culinary experiences or hipsters who exploit the revitalized Brooklyn foodscape, they both use food and foodways to construct identities that differentiate them from the mundane or the mainstream; however, they each do this in ways that erase the class privilege that makes it possible.

This is not to say that there is no self-awareness with regard to class privilege. When we visited Stinky Brooklyn, we encountered a fully branded experience where Stinky Brooklyn T-shirts and other souvenirs accompanied the seemingly endless array of international cheeses, the prosciutto and Ibérico ham bar, and the made-to-order gourmet sandwiches. Here—like at By Brooklyn, a gift and souvenir shop that boasts how all of its products are manufactured in Brooklyn—it is Brooklyn (arguably more an imagined than geographic space) that is marketed to locals and tourists alike. But while By Brooklyn is more overt in its marketing of Brooklyn as the source of the various *tchotchkes* for sale, Stinky Brooklyn markets a culinary experience that—while international in its reach—is very much identified with the local culinary scene. Both its products and its *mise-en-scène* market a culinary experience that is tied to place, namely Brooklyn's revitalized foodscape. And, as such, it can command top dollar from consumers eager to identify themselves with this cultural and culinary landscape. Matt Levy articulated this connection perfectly as we left the shop, and he justified paying ten dollars for a "hunk of cheese" by explaining what is really being purchased—a curated experience that will take shape as he and his wife enjoy this pricey cheese with the proper drink (they have started making their own bitters) and some good olives. Such investments (of time and money) are necessary, he argued, in an age when we are all so distracted by technology. Of course, they are also experiences available only to those who can afford them.

Affordability was also a consideration when we visited Brooklyn Kitchen in East Williamsburg, the brainchild of husband-and-wife co-owners Harry Rosenblum and Taylor Erkkinen. The business was started, Erkkinen explained, as part of her growing-up process as her lifestyle changed from closing down bars to hosting dinner parties (T. Erkkinen, personal communication, March 29, 2014). Consequently, she needed the knowledge and the resources to facilitate that change and had trouble finding either in her Brooklyn neighborhood. So, she and her husband hit upon the idea of opening a young, urban kitchen store. Two days after opening, in November 2006, a no-knead bread recipe appeared in the *New York Times*, which immediately sent oodles of customers through their door looking for the large Dutch oven called for in the recipe. They soon realized that the lifestyle changes they had been experiencing were apparently shared by many of their Brooklyn neighbors, which made the timing perfect for a store that offered cooking classes (learning basic knife skills, making homemade pasta, etc.), sold basic equipment, and would eventually carry food as well. Their success made expansion possible, and in 2010 they teamed up with The Meat Hook butchers to open their current location, which fully integrates two separate spaces for cooking classes, two floors of kitchen equipment (everything from cheese knives to KitchenAid stand mixers), a fully functioning butcher shop, and an extensive grocery store that combines local artisanal products, fresh fruits and vegetables, and a range of packaged foods. While the success of Brooklyn Kitchen attests to the fact that Rosenblum and Erkkinen were correct to see their own changing lifestyle as representative of similar changes being experienced by their neighbors, it also underscores shifting ethnic and class demographics in this and other Brooklyn neighborhoods that enabled not only their success, but also that of the Brooklyn culinary renaissance.

The relationship between these demographic changes and the success of Brooklyn's culinary renaissance is particularly striking when read in relation to this new generation of food artisans' tendency to invoke ethnic nostalgia as a motivation for their own culinary forays. Jennifer Causey encountered such claims when she interviewed Alison and Matt Robicelli, owners of Robicelli's Bakery in Sunset Park. In emphasizing the importance of their own ethnic roots to their work, they simultaneously highlight Brooklyn's rich ethnic diversity and its contribution to their success, explaining that "even though we were raised in our ethnic culture we were also raised in the Irish, Chinese, Polish, Mexican, Russian, Haitian, etc. cultures of our friends. The 'authentic' ethnic experiences that so many people travel the world looking for were in the houses of our neighbors.

We grew up in a world where we were always exposed to and trying new things. In our cooking, we draw from global influences with ease, because our childhoods were filled with these flavors in the same way most Americans grow up with hot dogs and hamburgers" (Causey, 2013, p. 163).

In this formulation, Brooklyn serves as the crucial site at which the local and the global converge, offering the Robicellis a childhood that is at once deeply rooted in their own ethnicity and also the basis for their ability to offer "authentic" ethnic experiences to customers seeking to connect to a real or imagined ethnic past. Michael Lomonaco strikes a similar note in his foreword to *Gastropolis: Food and New York City*, in which he contends that his unique perspective on food is rooted in his Brooklyn childhood, "where the fusion of great Jewish delicatessens and dairy stores, next to the Italian *salumeria* and *latticini*, down the block from the Mittel-European sweet shop that made ice cream and the market that sold fish brought fresh from the dock, and on whose side streets plied 1940s-era trucks from which was sold produce grown in the Hudson Valley conspired to make food the central focus of my life" (2009, p. x).

Though such nostalgia is often used to connect the current food boom in Brooklyn to its rich and diverse ethnic culinary past, it often fails to take into account the differences between the working-class, first- and second-generation immigrants, who relied on these culinary traditions for their survival, and their third- and fourth-generation children and grandchildren, who embrace them in their quest to create identities as hipsters or food people, who are steeped in a kind of "authenticity" that separates them from the masses. In tracing the evolution of New York's Jewish food, for example, Jennifer Berg notes that it was during the postwar suburban exodus that "many people abandoned their lowly ethnic foods, favoring more mainstream fare" (Berg, 2009, p. 270). However, as the nostalgia craze that peaked in the 1990s brought many of their children and grandchildren back to New York City, they embraced certain "ethnic" foods but did so without necessarily connecting them to "their own specific group identity." Rather, such foods became part of the cultural and culinary identity of New York, which Mitchell Davis describes as an "obsession with the newest, the latest, and the trendiest restaurant, food, and other cultural products [which] is about staying ahead of everyone else" (2009, p. 295). Thus Brooklyn's food artisans, like other savvy New York food people, may look back nostalgically to their immigrant ancestors for a sense of authenticity in their work, but they must also recognize the extent to which the specific foods and foodways they embrace have themselves become commodities in today's food culture.

From this perspective, the embracing of ethnic foodways, both by hipsters and by food people, can be read as conflicting with the very immigrant, ethnic communities to which they each stake a claim in their quest for authenticity. This was evident during our visit to Shelsky's Smoked Fish, when Peter Shelsky insisted that we stop at a Yemeni restaurant just around the corner on Court Street, explaining, "It's just like being in Yemen. They don't speak English. The food is *insane.* Just ask for this"—at which point he wrote "haneez with fattah" on a piece of paper. Even as Shelsky traced the roots of his own business to his immigrant ancestors, he made it clear that he did not view them or himself as the ethnic "others" against whom middle-class American consumers constructed their mainstream identities. Instead, he situated the Yemeni restaurant as an "authentic" site of such "otherness." In doing so, he exposed the fault line between the nostalgic embrace of an ethnic heritage that informs his and so many of the other culinary hotspots that have created the recent food boom in Brooklyn and a very different ethic culinary experience that is struggling to survive amid this increasingly gentrified landscape.

As we enjoyed our *haneez* with *fattah*, we speculated with Matt Levy as to how this Yemeni restaurant survives on a now-thriving Court Street, just across from that foodie bastion, Trader Joe's. Levy speculates about the ownership of the building that houses the restaurant (they must own it rather than rent the space), underscoring the difficulty many businesses have had remaining in neighborhoods that have emerged as new foodie enclaves. Although the proprietors of the Yemeni restaurant may have the security of owning their real estate, Levy recounts other businesses that have closed or relocated in the wake of rising rents. And for other businesses, survival comes from expanding product lines to attract more customers while simultaneously marketing an "authentic" experience of "otherness" to them.

This is what we encountered in Sahadi's, a Lebanese-owned specialty store founded in 1948. While they have recently expanded and continue to enjoy robust crowds, we were curious about the extent to which this success depended on balancing their exoticness as a "specialty and fine foods store" with the need to reach a wider customer base in search of an imagined culinary experience of otherness. Such a balancing act—and the reasons for it—are made explicit on the virtual "store tour" on their website. On the one hand, they highlight the Middle Eastern roots of the store, emphasizing the ways in which its current iteration maintains these traditions:

> When you first walk through the door, the aroma of fresh spices and Mediterranean cooking envelopes your senses and draws you into the bustling atmosphere of this iconic Middle Eastern grocery store. With bins of fine grains and exquisite spices, bulk containers of imported olives, nuts, and dried fruits, old fashioned barrels of coffee beans and the simple ticket dispenser from which customers must "take a number" during crowded times—this feels like an old-world place—and it is. Sahadi's has been doing business in Brooklyn for over 65 years with the original Sahadi's established in Manhattan in 1898. (Sahadi's, 2014)

On the other hand, even as it seeks to reinforce its Middle Eastern identity, the virtual tour acknowledges that Sahadi's is dealing with a very different customer base as it anchors a Middle Eastern shopping district that "endures where there are no Middle Eastern residents to speak of." Thus, its appeal shifts from serving a specific ethnic community to meeting the needs of a range of culinary enthusiasts, "from the gourmet chef, to the adventurous home 'experimenter.'" To do so, the market both emphasizes and waters down its ethnic identity, catering to anyone in search of culinary "otherness."

For example, the website boasts that the cheese department features "some Lebanese and Syrian cheeses you're not likely to find anywhere else in town" while also promising "a full line of specialty cheeses from around the world including farmhouse cheddar, a multitude of blues, Swiss and soft ripened, as well as five types of Feta." Such broad internationalism replaces ethnic specificity, as evidenced by how this department "also encompasses smoked fish products, frozen and refrigerated pastas and a full line of pates [*sic*] and sausages." Such balancing of ethnic specificity and generic otherness is also highlighted as the website summarizes the store's packaged items: "There's cricri, exotic herbs like sumac and malhab (made from the inside of cherry pits and used to flavor the Syrian string cheese), and mloukhiyeh (a forbidding spinach-like vegetable used in soups and gravies). There's Afghan bread as big as a pillow case and Turkish delight studded with pistachios. There is a full line of Middle Eastern products of course, as well as products from all over the globe" (Sahadi's, 2014). If you are looking for a truly exotic culinary experience, step right up! No need to wander through the diverse ethnic neighborhoods of New York City; come to Sahadi's, where today's food people can find "products from all over the globe" under one roof. Although such marketing may allow Sahadi's to continue to thrive in a neighborhood that is devoid of Middle Easterners, it also underscores the problematic impact of the Brooklyn culinary renaissance on those ethnic, immigrant communities that fall outside its reconceptualization of the "new" Brooklyn foodscape.

Of course, such cultural and culinary negotiations within and among ethnic immigrant communities are not a new phenomenon. Smith documents: "It isn't just that immigrant groups have moved into New York and started selling traditional or modified culinary treats from their homelands. The city creates opportunities for groups to interact in ways that they might never have in their points of origin. Groups and individuals have learned from others, and what's frequently offered on restaurant menus or on the family table show signs of these interconnections. Although many restaurants purport to offer 'authentic' foods and beverages (and some do), it is more common that immigrants have added nontraditional styles and dishes to their culinary repertoires" (Smith, 2014, p. 68). Although such cross-cultural interactions have surely yielded culinary innovations, they also reveal tensions between those who seek to maintain their unique heritage within their ethnic communities and those who romanticize such histories in constructing their own cultural and culinary narratives. As for the young entrepreneurs who are at the forefront of Brooklyn's culinary renaissance, class privilege allows them to embrace both the hipster and the foodie identities that are sustaining it while presuming a connection to Brooklyn's ethnic immigrant past, even as they contribute to the process of gentrification that threatens its present and future.

It is this very confrontation that seems to be playing out in efforts to revitalize the Moore Street Market in East Williamsburg. Opened in 1941, this was one of the many indoor markets constructed throughout New York City when it was decided to move the ever-growing number of food pushcart vendors off the streets. Located just off of Graham Street, also known as the Avenue of Puerto Rico, the market boasts a long history in this ethnic neighborhood. One Yelp reviewer explains that *La Marqueta* is "the place to be for everything Puerto Rico without leaving New York. If you are curious or are familiar with Puerto Rican Products, culture and cuisine, this is the place for you. Located in Brooklyn, who knew that a shortcut to Puerto Rico existed right in our backyard." The reviewer goes on to talk about the food that is available at the market: "Hot Food [*sic*] is also served here, you are able to sit down and enjoy some typical Puerto Rican dishes such as: Alcapurrias, Pasteles, Sancocho, Rice and Beans etc. Pasteles and desserts are made fresh on the premises" (Leo L., 2012). From the stall that sells herbal remedies, oils, and religious artifacts to the woman who makes homemade pasteles, to the owner of the local record store who sells CDs and DVDs to Ramonita's, the sit-down café where you can sample many ethnic delicacies, *La Marqueta* plays a central role in this community as a site where cultural and culinary traditions are maintained.

At the same time, the current effort to revitalize the Moore Street Market reveals the extent to which doing so will require a delicate negotiation between maintaining its traditional ethnic identity and marketing it to a broader base of customers. Part of this negotiation revolves around whether the market can rely on local residents to sustain its historical role as a center of ethnic cultural practices and foodways or whether it can only survive by reaching out to new customers, including those who may be drawn to it for a more commodified experience of otherness or who may need other kinds of products and services to make it a viable business. As the hipster food scene that has taken root in Williamsburg continues to flourish and possibly expand, one could imagine the market evolving in a very different direction. Interestingly, our tour guide shared the story of a boutique cupcake stall that had opened in the market, only to close a short time later when it became clear that the market's current clientele would not pay three dollars for a cupcake. At the same time, one of the newest stalls is Eden's Organic Natural Juices and Bubble Teas. Owned and operated by a local resident who has recently embraced healthier food practices, the stall may prove to be an interesting barometer of the market's future. As the owner talks about his decision to open the stall, our tour guide suggests that he might be more successful than the cupcake purveyors because the owner grew up in the neighborhood and may be able to link his organic natural juices to *aguas frescas*, traditional Hispanic juice-based beverages. In other words, he may be able to strike the right balance between attracting two different and quite distinct audiences: those who come to the market to keep their ethnic foodways alive, and those who seek an "exotic" cultural and culinary experience but who would also welcome the opportunity to indulge in healthy, organic juice. Although the owner seems confident that his neighbors are ready to embrace the healthier foods that he offers at his stall, when our tour guide asks if we can sample something, the owner's wife says that she will bring us a Filipino specialty, deep-fried plantains. No healthy organic juice on this immigrant food tour!

Conclusions and Observations

The question of the direction that the Moore Street Market's revitalization will take relates directly to the larger question of the cultural and culinary future of Brooklyn. As this chapter reveals, the history of this borough is very much intertwined with its ethnic immigrant populations and the foods and foodways that have sustained them. At the same time, these ethnic foodways are playing an interesting role in the culinary renaissance taking place across many Brooklyn neighborhoods. As foodies and hipsters compete for culinary credibility, they do so by embracing food practices that have sustained ethnic immigrant communities for generations and by tracing their own culinary values and priorities to their real and imagined immigrant ancestors.

Thus, place provides the landscape upon which hipsters and food people both romanticize their ethnic immigrant "Other" and stake claims to their foods and foodways while participating in a process of gentrification that threatens the viability of these very ethnic immigrant neighborhoods. Future research about the quest for distinction both by food people and by their hipster others must offer a structural analysis of how this underlying process of disinvestment, gentrification, and reinvestment affects the Brooklyn immigrant Other.

This chapter aimed to explore the social, political, and ideological power of food and foodways in a particular place, but one can extrapolate the ideas presented here to other cities as well. Brooklyn is just one of many urban spaces in the United States where the tensions among foodies, hipsters, and ethnic Others arise. Foodies and hipsters find new ways to signify their status against the backdrop of a renewed interest in food and increasing property values. Although this has played out in interesting ways in Brooklyn, it is a variation on a theme that one might expect to find in any urban space marked by rapid gentrification and a residential influx of young, educated people that displaces or threatens ethnic Others.

References

Bell, D. (2002). Fragments for a new urban culinary geography. *Journal for the Study of Food and Society*, *6*(1), 10–21.

Berg, J. (2009). From the big bagel to the big roti? The evolution of New York City's Jewish food icons. In A. Hauck-Lawson & J. Deutsch (Eds.), *Gastropolis: Food and New York City* (pp. 252–273). New York, NY: Columbia University Press.

Causey, J. (2013). *Brooklyn makers: Food, design, craft, and other scenes from the tactile life.* New York, NY: Princeton Architectural Press.

Clayton, J. (2010). Vampires of Lima. In M. Greif, K. Ross, & D. Tortorici (Eds.), *What was the hipster? A sociological investigation* (pp. 24–30). New York, NY: n+1 Foundation.

Davis, B. (2013). *9.5 theses on art and class.* Chicago, IL: Haymarket Books.

Davis, M. (2009). Eating out, eating American: New York restaurant dining and identity. In A. Hauck-Lawson & J. Deutsch (Eds.), *Gastropolis: Food and New York City* (pp. 293–307). New York, NY: Columbia University Press.

Galluzzo, A. (2013). The "fucking hipster" show: Mocking hipsters in the service of capital. *Jacobin: A Magazine of Culture and Polemic.* Retrieved from https://www.jacobinmag.com/2013/05/the-fucking-hipster-show/

Greif, M. (2010a). Positions. In M. Greif, K. Ross, & D. Tortorici (Eds.), *What was the hipster? A sociological investigation* (pp. 4–13). New York, NY: n+1 Foundation.

Greif, M. (2010b). Epitaph for the white hipster. In M. Greif, K. Ross, & D. Tortorici (Eds.), *What was the hipster? A sociological investigation* (pp. 136–137). New York, NY: n+1 Foundation

Grimes, W. (2009). *Appetite city: A culinary history of New York.* New York, NY: North Point Press.

Johnston, J., & Baumann, S. (2010). *Foodies: Democracy and distinction in the gourmet foodscape.* New York, NY: Routledge.

Lanham, R. (2002). *The hipster handbook.* New York, NY: Anchor.

Leo L. (2012). Moore Street market. Retrieved from http://www.yelp.com/biz/moore-street-market-brooklyn-2

Lomonaco, M. (2009). Foreword. In A. Hauck-Lawson & J. Deutsch (Eds.), *Gastropolis: Food and New York City* (pp. ix–xi). New York, NY: Columbia University Press.

Mendelson, A. (2009). The Lenapes: In search of pre-European foodways in the greater New York region. In A. Hauck-Lawson & J. Deutsch (Eds.), *Gastropolis: Food and New York City* (pp. 15–33). New York, NY: Columbia University Press.

Naccarato, P., & LeBesco, K. (2012). *Culinary capital.* London: Berg.

Parasecoli, F. (2009). The chefs, the entrepreneurs, and their patrons: The avant-garde food scene in New York City. In A. Hauck-Lawson & J. Deutsch (Eds.), *Gastropolis: Food and New York City* (pp. 116–131). New York, NY: Columbia University Press.

Sahadi's. (2014). Store tour. Retrieved from http://www.sahadis.com/store-tour/

Santloffer, J. (2009). Asphalt *terroir.* In A. Hauck-Lawson & J. Deutsch (Eds.), *Gastropolis: Food and New York City* (pp. 174–194). New York, NY: Columbia University Press.

Shelsky's. (2014). About Shelsky's of Brooklyn. Retrieved from http://www.shelskys.com/about

Smith, A. F. (2014). *New York: A food biography.* Lanham, MD: Rowman & Littlefield.

Part III

The Context of Power and Inequality

Chapter 7

The Political Economy of Food and Agriculture in the United States and Russia

Susanne A. Wengle

Introduction

Most modern states take a keen interest in food production and farming. Governments pursue a variety of aims in this realm—from ensuring food supply and stabilizing farm income or consumer prices to protecting national or regional foodsheds. Multiple policy regimes regulate agriculture and food production, fundamentally shaping domestic and international markets for food and agricultural commodities. Though many political goals remain quite similar over time and across different economies, how policy regimes function and what their effects are have varied widely. Even one type of policy in one economy rarely affects all constituents equally. Policies redistribute resources via public programs or via the price and supply of commodities, variously shaping incentives for producers and consumers in different sites. A place-sensitive approach to studying the political economy of food produces analytical maps that draw attention to how policy regimes interact with particular locales. Such an approach could grasp, for example, how agricultural policies affect rural and urban constituencies differently. Urban constituencies tend to benefit from policies that address the affordability, quality, and reliability of food and the cost of producer-support programs to taxpayers. Food stamps are an entitlement program that has traditionally catered to urban constituencies. Rural communities, by contrast, generally benefit from and favor policies that counteract the volatility of commodity prices and lift farm-gate-commodity prices. The urban/rural distinction is only the most obvious fault line in a place-sensitive map of food politics. Another way of drawing up a map of food politics would take into account types of commodities and the places in which they are grown, or the scale of production. Each of these factors interacts differently with various policies: programs that help large-scale producers of commodity crops in a country's agricultural heartland can be irrelevant or can even work against farmers who pasture livestock in more marginal areas, for example.

Established debates in political economy of agriculture tend to examine the policy-making process of market regulations. They examine, for example, how agricultural policies reflect the ability of different interests to influence policy-making—be that through access to the policy-making process or by shaping public and expert discourse (Hansen, 1991; Sheingate, 2003). These remain important questions. More recently, a growing body of research in political science, sociology, and public policy has started to examine policies and market regulations as ongoing processes in which the regulating state interacts with communities and stakeholders (Abers & Keck, 2013; Espach, 2009; Berk, Galvan, & Hattam, 2013; Overdevest, 2010; Sabel & Zeitlin, 2010). This scholarship no longer thinks of the regulating state as relatively autonomous in its ability to implement policies and to "command and control" the economy as it sees fit. Instead, these studies (one could group and name them as complex governance approaches) view policy regimes as complex regulatory arrangements that arise from particular political constellations but at the same time interact and shape political and market positions, hence interests themselves. Regulatory regimes, then, both rely on and generate political interests.

For a place-based perspective, this notion of governance is interesting, because it disaggregates how policies function across different localities and communities. It can draw attention to the politics of regulation as site- or place-specific interactions between policy regimes and the regulated communities. We will see below that in the two countries examined here, the United States and the Soviet Union, benefits of various policies were unequally distributed across space and different commodity types. In very broad terms, policies benefit a relatively small number of large farms, and they have been concentrated in a few states in the United States and in the fertile black-earth region in Russia. This is important, because there is strong evidence from research in rural sociology that small and medium-sized, independently owned farms tend to be an asset to rural communities (Lyson, Stevenson, & Welsh, 2008), and from debates on rural development that small farms alleviate rural poverty (Deininger & Byerlee,

2011). To the extent that large farms were more likely to be beneficiaries of various types of policy regimes, dominant policy regimes have not fostered forms of production that these debates identified as particularly beneficial for local communities, and they have *not* had positive effects on rural communities and have been *less* effective outside the agricultural heartland in both countries. While the scope of this chapter does not allow an in-depth map of how polities interact with particular types of producers in particular places, the aim here is to demonstrate the relevance of place-based research on the community linkages of different policy regimes and to show where opportunities for this type of research exist.

The theoretical aims of the chapter then are to sketch the contours of a place-sensitive approach to the political economy of agriculture and to outline how such a perspective would assess a set of important policies in the agriculture and food sectors. Empirically, the chapter documents how three important policy areas—subsidies, land-use policies, and trade policies—function differently across polities and locales even if they are motivated by similar rationales. Specifically, what follows is an overview of the efforts by the United States and the Soviet Union (and the Russian Federation after 1991) to regulate agriculture and food production in the postwar period. A comparison of the food and agriculture politics in these two polities and economies is interesting because their respective agricultural sectors were intertwined in many ways. They were linked by the competition over yields and technologies that the Cold War entailed and also increasingly by the food system's *internationalization*. Entangled in this competition, the two countries pursued similar aims. At the same time, policies in one country mirrored developments in the other. In a very short preview—the Soviet Union was struggling with shortages, while the United States had to deal with oversupply. As US agricultural exports soared, the Soviet Union became increasingly import-dependent by the 1980s and the 1990s.

In sum, this chapter suggests that the political economy of food and agriculture could produce maps that analyze and explain community linkages of policy regimes. The next section introduces agricultural policies in the United States and in Russia, drawing attention to similarities and differences in what these policies aimed to achieve and how they functioned. The bulk of the chapter comprises the third section, which introduces three policy regimes in each of the two countries: (i) a variety of subsidies that direct public resources to farmers, (ii) land-use policies that either expand or contract sown acreage, and (iii) trade policies, such as export promotion and import restrictions. The conclusion draws on this discussion to restate what can be gained from a place-sensitive approach to political economy.

Agricultural Policies in the United States, the Soviet Union, and the Russian Federation

The United States and the Soviet Union are usually thought to be more different than similar. One of the principal differences between the two economies was their respective economic system, and indeed capitalist agriculture defines the American food system, whereas the Soviet Union relied on state planning. The role of the Soviet state in agricultural production was large and direct: it owned farms, it controlled input and output prices as well as farm wages, it made investment decisions, and decided what to plant and where, to name a few differences. The American state influenced production and markets indirectly, via policy regimes that shaped relative prices and incentives. At the same time, despite this difference, there are interesting parallels between the United States and the Soviet Union. For once, for much of the twentieth century, both the United States and the Soviet Union enthusiastically endorsed the principles of industrial agriculture, and the goals they pursued were remarkably similar (Fitzgerald, 2003). Moreover, in both countries, agricultural policies reflected shifting urban-rural bargains. Initially as part of the war on poverty, US farm bills entailed benefits for urban constituencies through growing food stamp programs after the mid-1960s. In the Soviet Union, a number of mechanisms redistributed resources from the countryside to the city, principally through low commodity prices and industrial wages that were set far higher than agricultural wages. The terms of this redistribution varied, and in both economies they shifted considerably over the years. But both were ways of subsidizing urban consumption; in the United States, it came to be thought of as the urban-rural logroll and, in the Soviet Union, as the dictatorship of the proletariat.

A further interesting similarity relates to the size of farms: despite the differences in the way food was produced in the United States and the Soviet Union, in both countries, agricultural production became increasingly bifurcated—into very large and very small units of production. In both countries the "middle" is either disappearing or has virtually disappeared. The postwar trajectory of the food system in the United States charted an essentially bifurcated path. The North American (and increasingly also the global food system) became dominated by large-scale, consolidated agri-food corporations that control supply chains (Lyson et al., 2008; MacDonald, Korb, & Hoppe, 2013). By and large, this type of farming and food production is premised on reducing production costs, principally through economies of scale and technological innovation. Though yields have increased greatly, this mode of production has also had the tendency to externalize costs, both on the environment, through pollution and soil depletion, and on human health. A countervailing trend has been the growing popularity of alternative forms of farming, of producing and marketing food, which are distinct from industrialized farming through their emphasis on sustainability. Part of this trend is the (re)building of short and local supply chains that sell directly to consumers, traditionally via farmers markets and community-supported agriculture (CSA), but increasingly also through a variety of new organizational forms, such as coops and food hubs (Goodman, Dupuis, & Goodman, 2012). In the United States, these trends have contributed to what many have called the disappearance of the agriculture of the middle (Lyson et al., 2008). Midsized, family owned farms that have traditionally relied neither on contractual relations with highly integrated, global commodity chains nor on direct marketing have found it very difficult to remain commercially viable

The structural changes of farming in the Soviet Union in the twentieth century were marked by different events, though the similarities in terms of the bifurcated, or bimodal farm structure are striking. After the forced collectivization of agriculture and the creation of the *sovkhozy* (state farms) and *kolkhzy* (collective farms) under Stalin, the socialized, centralized, and planned agricultural production defined the Soviet food system for decades. By all accounts, Soviet agriculture was in permanent crisis, plagued by chronic shortages and gross inefficiencies. Although aggregate output increased in the postwar period, the collective farms could not keep up with the demand for grains and other foodstuffs, despite the ever-larger capital, land, and labor resources that were allocated to the collective sector. Alongside these large-scale, notoriously inefficient production units, *lichnoye podsobnoye khozyaystvo* (small-scale private farming) existed as a second characteristic form of agricultural production (Ioffe, Nefedova, & Zaslavsky, 2006; Pallot & Nefedova, 2007). These were private plots that rural workers were allowed to farm and the suburban plots for urban residents. Production and distribution of foodstuffs produced on small private plots happened entirely in the informal economy. The produce, eggs, dairy, and meat originating in this sector were largely consumed by the producers' family members or exchanged or bartered in local networks.

Throughout the Soviet period, private plots remained absolutely essential contributors to the food supply; since the collapse of the Soviet Union they have acted as important safety nets for newly vulnerable populations (Hedlund, 1989; Ioffe et al., 2006; Pallot & Nefedova, 2007; Wegren, 1998). As in the United States, though with a very different trajectory, there are virtually no medium-sized, owner-operated farms in contemporary Russia. Collectivization had evicted, even eradicated *kulaks* (private farmers) in the 1930s. Post-Soviet land reforms sought to re-create family farms. The creation of private agriculture faced a number of obstacles and essentially failed. Farmers found that despite the legislation in their favor, in reality, land was difficult to obtain (especially in the first few post-Soviet years). Even if farmers managed to officially secure private land, they were often unable to obtain credit to update machinery. Private farms also remained tethered and dependent on the collective sector for various inputs such as water, electricity, and gas and had difficulty obtaining credit to acquire machinery or livestock. The most successful private and collective farms entered into contractual relations with large, vertically integrated food-processing companies (Ioffe et al., 2006).

As stated earlier, a number of rural sociologists believe that the agriculture of the middle has unique community linkages; this means that the bifurcated modes of production in which the "middle" is increasingly squeezed (as in the United States) or is failing to reemerge (Russia) are particularly important for a discussion on the community linkages of policies. How policy regimes function in these contexts and to what extent they are implicated in creating bifurcated structures of production are open questions. The discussion of policy regimes below will begin to table some of the evidence we have to answer them.

Policy Regimes

While agricultural policies pursue long-term political, economic, and social objectives, they are also always attempts to address chronic and acute problems in the way food provisioning functions. While the United States and the Soviet Union pursued similar goals—attempting to modernize and rationalize by seeking ways to boost efficiency and productivity—the problems policy makers faced, the obstacles in the way of these goals, were different in the United States and Russia. In fact, they were often virtually mirror images of each other. One of the most salient problems that US agricultural policies sought to address postwar was the problem of chronic disequilibrium between the demand and the supply of agricultural products. Technological advances in farm machinery, agricultural chemicals, and hybrid seeds led to a rapid increase of yields in the postwar period and hence a burgeoning supply of farm products. Given that demand for agricultural products is relatively inelastic—save for population growth, there is only so much food a society consumes—the US food system increasingly overproduced agricultural commodities. Over-supply threatened to create a reinforcing spiral of falling prices, inducing farmers to plant even more, which in turn exacerbated the problem of supply-demand mismatches. There was often little agreement as to what could be done to address this problem, and policies that were devised over time were derived from different approaches to dealing with it (Ingersent & Rayner, 1999). In very broad terms, successive administrations authored policies that either principally addressed the problem of farm incomes or sought to find ways of dealing with the supply side of the equation by managing the commodity surplus. Various types of subsidies—direct and deficiency payments and subsidized farm insurance—fell into the former category. Land-reserve programs, export promotion, and the food-stamp program represent the latter.

The problems in the agriculture and food sectors that successive Soviet policy makers struggled to address looked quite different. Quite unlike the capitalist system, the Soviet planned economy was unable to produce an abundance of commodities; and shortages were the overarching problem that communist leaders contended with decade after decade. The Soviet Union was far less successful in increasing yields and efficiency. Indeed, the private plots that rural residents maintained were the most productive part of Soviet agriculture. Although these small plots were an important contribution to the country's food supply, the vast majority of rural land was farmed in the collective sector. The collective sector's productivity was shaped by wage and commodity pricing policies that privileged industrial labor and urban interests. In general terms, prices for agricultural commodities were set at low levels to keep food affordable for urban consumers. Similarly, rural wages were kept low, especially during the first part of the postwar period, when wage policies were meant to equalize rural labor, and specialists and farm managers earned only a little more than farmworkers (Wegren, 1998). Price and wage policies also ended up rewarding high-cost producers, did little to stimulate efficiency and yields, while contributing to the pervasive food shortages.

The sections that follow will examine three policy areas in the United States and the Soviet Union in greater detail; in each, we will discuss whether and how they interacted with different types of farming operations in different places. If different forms of production and ownership are linked to local communities in various ways, as a growing body of literature suggests, it is important to examine whether policies foster a particular type of production.

Subsidies: Price Support, Subsidized Insurance, and Other State-Directed Resources

Agricultural subsidies have been one of the principal policy instruments in the US agriculture sector, as in most other advanced industrialized countries (OECD, 2013). *Subsidies* is an umbrella term for a variety of safety nets that support farmers, shielding them from fluctuating commodity prices and the risks of crop failure. The chief rationale for subsidies has been the stabilization of commodity prices and farm incomes. In the United States, they have traditionally been administered either as direct payments (compensations in case of falling prices), as subsidized loans, or as insurance schemes (in case of crop failure). Among the most controversial aspects of these programs was whether they should guarantee a fixed price, or whether target prices should fluctuate with market prices; the former guaranteed farm income and insulated against price volatility, while the latter was primarily meant to create an income safety net. In the 1960s, for example, wheat, feed grains, and cotton producers were eligible for programs that offered fixed support prices (Ingersent & Rayner, 1999). Insurance schemes date back to the early 1970s, when the government first committed to direct payments to cover losses due to natural causes that prevented planting or significantly lowered yields. By the early 1980s, disaster payments were replaced by subsidized all-risk crop insurance programs. As long as agricultural prices were relatively high, as they were for most of the seventies, the cost of these programs was defensible. When commodity prices fell, however, as they did in the eighties, the price tag of compensation programs soared, and political pressure to raise reference prices and price floors mounted. Since the nineties, direct payments have gradually been replaced with various types of crop insurance, which were meant more to insulate farmers against catastrophic shortfalls and less against market fluctuations. Various programs to compensate farmers for low market prices continued, in the dairy and sugar sectors for example, even as support prices were gradually lowered. The most recent farm bill, passed in 2014, continues this shift away from compensatory direct payments and toward new forms of subsidized crop insurance against crop and livestock losses.

The benefits of compensatory payments have been unequally distributed. In very broad terms, subsidies tend to benefit commodity crops; they go only to a small number of large farms, and they have been concentrated in a few states. Corn farmers were by far the largest recipients of federal subsidies, followed by wheat and corn growers (Environmental Working Group, 2012). Fruit and vegetable growers have historically not benefitted from federal subsidy and crop insurance programs (Sumner, 2008; UCS, 2012). A small number of large farm operations receive the bulk of the benefits. To take data on a recent period, for example, from 1995 to 2012, a majority of farms (62 percent) did not collect subsidy payments. Ten percent of farms collected most (75 percent) of all subsidies. Note that as subsidies are tied to production, larger farms that grow more tend to collect more—though subsidies make up a smaller share of overall income of these large operations, compared to smaller ones. A very small percentage of large farms received very large sums; in 2011, for example, twenty-six farms received insurance subsidies exceeding $1 million (Environmental Working Group, 2012). Federal subsidies go primarily to the Midwest and to the South, though various programs have different geographical concentrations. Direct payments dominate in the Corn Belt (to corn and soy growers) and the Mississippi Delta and Gulf Coast (cotton and rice growers), while insurance indemnity payments are the most important subsidy in the wheat-growing areas of the Great Plains (USDA Economic Research Service—AIS90). Texas, Iowa, and Illinois lead in terms of overall federal subsidy payments.

In the Soviet Union, there was no internal market for agricultural commodities, so farmers did not need to be shielded from price volatility. The state determined rural wages and procurement prices for farm output and throughout the state procurement system. Purchasing prices for agricultural commodities, together with capital investment and labor allocation, were the central policy tools that the Soviet leadership used to achieve its political and economic goals. During the first postwar decade, agriculture received far less state investment than did industry, as noted above, as procurement prices for agricultural commodities and rural wages were low, stagnant, or even decreased (Ioffe et al., 2006; Wegren, 1998). The state also allocated food-related privileges to party elites, reserving access to specialty stores and restaurants for them. These policies were a manifestation of the communist leadership's explicit antirural bias. (At the same time, rural residents as a whole suffered far less from food shortages because they had easier access to food grown on private plots.) By the 1970s, attempts to bring more land under cultivation (see below) had largely reached their limits; and the focus of farm policy turned to mechanization and technological updates, which meant that capital allocations along with procurement prices were raised. While more resources flowed to collective farms, Brezhnev also pursued the goal of rural egalitarianism. This, in turn, meant that less productive farms and farms located in relatively less fertile zones were eligible for *nadabavki* (a type of subsidy), meaning they received higher purchasing prices than did strong farms in the agricultural heartland. Though surging capital allocations in the 1970s led to a tremendous growth in agriculture's capital stock, they were accompanied by far smaller increases in output—most Soviet farms remained inefficient. By the time Gorbachev came to power, almost half (48 percent) of farms were either barely profitable or not profitable, hence in the category eligible for higher procurement prices, the *nadabavki* (Wegren, 1998). Rural wages increased under Gorbachev, but they remained well below industrial wages.

In 1991, after the collapse of the Soviet Union, the paradigm of state support changed. With the sudden and dramatic collapse of state support and the state procurement system, many rural regions were struggling to recreate supply chain relationships in the new market system; and the overall output of Russian agriculture declined steeply during the 1990s (Wegren, 1998). As the state procurement system crumbled, far fewer resources flowed to the countryside, though the state also subsidized agricultural inputs through in-kind subsidies (for example, via low-priced energy). State support for agriculture has recovered in the most recent decade; various subsidies to support the development of the livestock sector are a priority of the Russian government under Vladimir Putin (Vassilieva, 2012; Wegren, 1998). On the whole, rural regions located closer to urban centers and central black-earth regions were better able to recover and to transition to a marketized system, while in less fertile regions, arable land is abandoned because residents move to cities (Ioffe et al., 2006).

As in the United States, the effects of Soviet agricultural policies were unequally distributed. Soviet price, wage, and investment policies that channeled resources to the countryside were almost exclusively directed at the *kolkhozes* and *sovkhozes*; they did not directly flow to private plots. Yet, at the same time, the private plots often obtained much of their input via the collective sector—wages were partly paid with feed grains, private cattle grazed on collective farmland, or collective farm chiefs bartered other inputs for produce grown on tiny plots, for example. Labor on private plots was entirely voluntary, though also often made possible by public-sector employment (on the collective farm or as teachers) and pensions. Because of this symbiosis between the collective sector and the private plots, policies that directed resources toward the former usually ended up indirectly benefitting the latter. During certain periods of Soviet history, private plots flourished under these conditions. At other times, resources invested in this form of agriculture were drained because they were diverted toward fulfilling plan targets of collective and state farms. Private plot holders' rights were often tenuous and at the mercy of a variety of actors—including local authorities and collective farm managers (Pallot & Nefedova, 2007). During a campaign to surpass the United States in meat consumption in the late 1950s, for example, collective farm managers and local party officials pressured rural residents who held livestock on private plots to sell or hand over cows to the collective farms. The effect of this campaign on private livestock was that large numbers of livestock were prematurely slaughtered, devastating holdings for years, if not decades (Gorbachev, 1996). The small private plots continued to be important during the post-Soviet period. The economic crisis that accompanied the collapse of the planned economy meant that collective farms had difficulties locating inputs and selling output, which curtailed acreage and reduced livestock holdings. By contrast, the share of food produced on tiny private plots surged in the countryside and in rings around urban areas—in some years and for some commodities, accounting between 50 and 90 percent of production, depending on the commodity (Федеральная служба государственной статистики, 2000). At the same time, despite the importance of household farming, the state continued the Soviet-era policy of relying on this form of production while largely ignoring its needs and challenges. No policies were aimed directly at supporting this sector. Indeed, after 2000, a consolidation of agricultural land in very large, vertically integrated agriholdings took place (so-called NAOs, new agricultural operators). These new owners were often not interested in upholding the long-standing, informal agreements that allowed resources to flow from the collective land to private plots (Pallot & Nefedova, 2007; Nikulin, 2003).

Land-Use Policies

Land-use policies in the broadest terms are attempts to influence a farmer's decision regarding whether and how to farm the land. In the United States, successive policy initiatives sought ways to induce farmers to voluntarily idle acreage, as the specter haunting American farm income was agricultural surpluses. These programs, known as acreage reserve programs (ARPs), continued what had been called set-aside schemes, which dated back to the 1930s and were primarily aimed at preventing soil erosion. As yields and surpluses grew rapidly in the postwar decades, attempts to curtail grain production took on a central role in land-use policies. In the early 1960s, a number of programs sought to lower stocks of feed grains, paying farmers to either idle or divert land away from feed-grain production. Starting in the mid 1970s, a growing awareness of the environmental pollutants that are specific to agriculture, such as pesticides, plant nutrients, and animal waste, led to the coupling of set-aside programs with conservation aims. This meant, for example, that land in crop-diversion programs had to be put to very specific alternative uses (Ingersent & Rayner, 1999). The cost of set-aside and acreage limitation programs was typically borne by taxpayers, rather than passed on to consumers via higher prices, as concerned policy architects carefully monitored the situation to ensure that reduced supply would not increase commodity prices and spur inflation (Ingersent & Rayner, 1999). Set-aside and reserve programs addressed the problem of oversupply, but they were often politically unpopular. Opponents of supply-side management considered them interventionist and antimarket, and if applied widely, too expensive. Several programs in the 1980s were highly successful in limiting crop-land, but they resulted in record costs and proved to be politically unpopular for the Reagan administration (Ingersent & Rayner, 1999). Over the last few decades, land-use policies that are relevant for agriculture have moved further into the direction of, and have been replaced by, broader conservation aims. Programs provided assistance to farmers who committed to address water management and soil erosion ("Agricultural Management Assistance.") As suburban sprawl and commercial development increasingly threatened agricultural land, land-use programs also increasingly sought to preserve certain land for agricultural use, through easements that limited nonagricultural use, for example.

Acreage-reserve programs initially targeted commodity crops and feed grains. As land-use policies evolved over the years to serve conservation aims, the type of farms that were eligible as beneficiaries broadened. It seems, though, that many recipients of conservation programs were also recipients of crop subsidies; at least this was the case for the period between 1995 and 2012 (Environmental Working Group, 2012). Payments in the conservation programs were somewhat less concentrated than in the subsidy programs: the top 10 percent of recipients collected 59 percent of payments. Iowa, Texas, and Kansas are the three states that have received the largest payments under federal conservation programs during this time period.

In the Soviet Union, by contrast, land-use policies sought to increase cultivated land to deal with the problem of shortfalls. Two types of land-use policies were most influential: ambitious and strident state-led programs that sought to extend the acreage of the collective sector, on the one hand, and the gradual, incremental, and hesitant expansion of land allocated to the private plots. A gargantuan effort by Khrushchev in the 1950s—to render arable vast tracts of the Virgin Lands, most of them located in the Central Asian steppe, and to bring them into the service of food production—was the most spectacular program of the former kind. The sheer size of the land that was brought under cultivation was spectacular—by the early 1960s, an area the size of the entire cultivated area of Canada was added to the total sown land of the Soviet Union. Programs that addressed agricultural production on the smallest scale, on the other hand, namely, the policies toward land use for *lichnoye podsobnoye khozyaystvo* were torn between an ideological aversion to private landholding and an increasing realization that this form of production acted as an important supplier of food. Under Khrushchev, ideology prevailed; various policies undermined and damaged private plots. As Soviet collective farms continued to fall short of plan targets in the 1960s and 1970s, and while urbanization accelerated, Brezhnev adopted a more pragmatic stance toward private plots. He allowed larger plots to be farmed privately, and irrigation and livestock holdings were also once again permitted (though notably not encouraged actively). This policy change meant that private plots became more productive—an important achievement, given the general shortages that plagued the Soviet food system. Animal products were of particular concern to Brezhnev, who welcomed the growing share of meat, milk, and eggs produced on these tiny private plots. Gorbachev continued the policy of turning to personal plots to bolster food supply, though he also realized that they were very labor-intensive and could not ultimately solve the problem of Soviet agriculture. During the first post-Soviet decade, land was the single most important resource allocated by the state, and land was meant to be distributed to rural residents in the privatization of the collective farms in the 1990s. However, the creation of new private farms was initially hindered by a number of factors, including the unwillingness of collective farm managers to let go of the best land and by the absence of rural credit that would have made farming viable (Allina-Pisano, 2007; Wegren, 1998). During the 1990s, relatively few private farms were created, and collective farms remain the dominant ownership model. The farms that were created were most likely located in Russia's fertile regions—along the river Volga and in the North Caucasus region (Wegren, 1998). After 2000, rising global food-commodity prices and the active support of the Putin government contributed to the consolidation of landownership by the NAOs, who consolidated ownership of very large farms. Note that land consolidation was a geographically uneven outcome: we see aggressive land accumulation by NAOs in black-earth and Russia's southern regions, while in the northern regions, state policies did not address the difficulties of the rural population nor forestall land-abandonment and depopulation (Rylko, Jolly, Khramov, & Uzun, 2008).

Trade Policies

During the first two postwar decades, international trade in agricultural products was limited as countries protected their markets and suppliers in former colonies with a myriad of trade barriers. In the 1970s, agricultural goods increasingly became internationally traded commodities, as the demand for imported grain by the Soviet Union and by developing countries that were undergoing rapid urbanization soared. In the United States, the encouragement of exports and foreign aid was another major approach to dealing with the problem of agricultural surpluses; both served to divert surpluses to foreign markets. With highly productive farms, free trade for agricultural commodities became a US foreign-policy priority, and the country became the world's largest exporter of farm products in the postwar period. The Foreign Agricultural Service (FAS), created in 1930, and charged with promoting exports and coordinating food aid in the wake of the Second World War, played a significant role in this development. In the first postwar decades, the main export destinations were Western Europe and Asia; by the 1970s, grains increasingly went to the Soviet Union and Eastern Europe. As food aid, exports were an important channel for surplus disposal: for some years, around one-third of all agricultural exports took the form of nonmonetized food aid for "friendly" developing countries (Ingersent & Rayner, 1999). Though the effects of food aid on recipient countries were often problematic, at the time this was thought of as a successful combination of foreign policy objectives and domestic agricultural policy. Moreover, domestic critics of direct payments and subsidies for farmers embraced international trade and export promotion as a "free market" solution to the problem of domestic surpluses. Contra the free-market rhetoric, however, US exporters received assistance through various programs and export-related subsidies—support measures that have repeatedly been at the center of high-profile trade disputes. Whereas some measures were curtailed, following the Uruguay round of multilateral trade negotiations, others persisted. Trade promotion, export-credit guarantees, and other export-related subsidies were allowed to continue, as they were placed in a category deemed nondistorting (Ingersent & Rayner, 1999). Export subsidies and export promotion—currently known as foreign market development programs—remain a major element of US agricultural policy.

As with other types of subsidies, producers of certain types of commodities were more likely to benefit from various export subsidies. While the United States is a strong exporter in a number of commodities, including poultry, beef, and pork, field crops have traditionally dominated exports (soy, corn, wheat, and cotton). Among these, soy and its related products are by far the most important export commodity, making up 41 percent of US agricultural exports. Illinois, Iowa, and Minnesota are the leading soy exporting states (USDA Economic Research Service, 2013). The United States also has a long history of supporting dairy and cotton exporters, even though dairy does not top the list of exported agricultural commodities; generous export subsidies in both sectors have been exceptionally controversial in international trade disputes.

Starting in the early 1970s, successive Soviet governments turned to imports as a solution to the problem of shortages. Grain and dairy products were imported on a large scale to meet plan targets. Wheat and feed grains made up the bulk of Soviet imports from the United States, and they increased throughout the 1970s. Soviet imports of US grains were affected by the seesaw in the political relations between the two Cold War superpowers, and the United States imposed a trade embargo after the Soviet invasion of Afghanistan. As the Soviet Union shifted to Eastern European grain in the late 1980s, efforts by the Reagan administration to counterbalance the effect on US exporters included subsidies for grain exports destined for Russia (Becker, 1987). During the post-Soviet period, Russian trade policies shifted to curtailing imports and protecting domestic producers with tariffs. Imported meat, poultry, butter, and sugar, for example, were consistently subject to import duties, and on many occasions, particular products were banned on the rationale of food-safety concerns (Ioffe et al., 2006). By imposing import duties, the government has tried to foster domestic capacity in agriculture. Still concerned about the availability of animal products, the Russian state also has several programs to promote the domestic livestock sector, including the subsidizing of animal-breeding import and the construction of large-scale cattle ranches. It is likely that trade policies have affected Russian agriculture differently across the country, though more research is needed on this issue.

Uneven Effects of Policies

Both in the United States and in the Soviet Union, certain places and types of growers were much more likely to be beneficiaries of subsidy and insurance policies. In the United States, state support generally flowed to producers of a limited range of commodities—grains, oilseeds, cotton, sugar, and dairy products. As farms producing these crops also tended to be the largest, public support disproportionately went to very large and profitable growers and farm operators. Producers of other agricultural commodities—including meat processing (beef, pork, poultry), hay, fruits, tree nuts, and vegetables—received far less, even minimal government support (Sumner, 2008). And insofar as these commodities are grown in certain places, the policies interacted with the particular locales where these crops are grown—primarily in Illinois, Iowa, Texas, and Kansas, and the agricultural heartland of the Midwest and the Great Plains, more broadly. Notably, California, where much of the country's vegetables are grown, and the Northeast, with its large urban-population centers, rank far lower in state support compared to these regions. In the Soviet Union, support overwhelmingly flowed to the collective sector; unlike in the United States, collective farms *outside* the agricultural heartland were long the disproportionate beneficiaries of state resources. In the post-Soviet period, marginal regions suffered disproportionately as the state procurement system collapsed. Collective farms that were already unprofitable in the Soviet period barely managed to keep operating in the post-Soviet period, which meant that rural outmigration affected these regions particularly hard (Ioffe et al., 2006). After 2000, the NAOs in the black-earth and southern regions were the main beneficiaries of state support. In terms of the type of producers, much like the United States, Soviet and Russian farm policies clearly favored large producers; since the turn of the century, state policies were designed to support NAOs and at times were openly hostile toward household producers. Despite their importance, small private plots were, at best, residual beneficiaries of

Soviet and Russian rural policies; they largely thrived and/or struggled in the informal sector.

Conclusions and Observations

We know that the United States and the Soviet Union both created a bifurcated, or bimodal, system of production. At the end of the twentieth century, the agricultural sectors of both countries were dominated by a small number of large farms and a large number of small farms. In the United States today, the large farms produce the bulk of output, while small farms produce comparatively little. In the Soviet Union, small farms produce far more than their size and their labor-intensity suggest. In both countries, small farms are often not commercially viable, and farms survive only because farm income is supported by off-farm income. In both countries, the "agriculture of the middle" is either disappearing rapidly (in the United States), or failing to reappear (in post-Soviet Russia). Farms that make up these middling types are the most vulnerable in polarized markets—too small to compete in global (or even national) commodity markets but too big to survive on direct marketing alone (Kirschenmann, Stevenson, Buttel, Lyson, & Duffy, 2008). What we do not understand fully is how and to what extent agricultural policies have produced this outcome. This is an important question, because medium-sized, owner-operated farms may bring advantages that neither of the other two organizational forms can provide. Medium-sized farms uphold diversified production. Unlike farms that are reliant on mono-crops used primarily in processed food, medium-sized farms have the flexibility to respond to consumer demand for high-quality, fresh food. But they can farm larger acreages and produce more food than small direct-marketing farms that need to juggle the growing, preparing, marketing, and delivering of food.

The fate of the agriculture of the middle is also relevant for a place-sensitive analysis of agricultural policies according to a body of research on the community linkages of different types of farming. Known as the debate on the agriculture of the middle, this research argues that different forms of agricultural production relate differently to the communities. In a nutshell, the argument is that small- and medium-sized, independently and family-owned farms tend to make decisions with the long-term sustainability of the farm in mind. Acting as "stewards of the land" (Kirschenmann et al., 2008, pg. 9), they are less likely to externalize the cost of production on the environmental and rural communities. Large-scale facilities that rely on decreasing the cost of production through economies of scale, by contrast, are more likely to produce waste, pay low wages, and produce for distant markets. Kirschenmann et al. summarize the benefits of the farming of the middle: “In addition to managing the farm for profitability, most [independent, family-owned] farmers also made decisions that assured the survival of the farm in its particular community so that it could be passed on to future generations in good health,” though they acknowledge that “small, independent farms have always managed to prevent soil loss, protect water quality, [and] maintain vibrant communities” (p. 9). The writings on “civic agriculture” theorize the embeddedness of a particular type of farming with local communities even more explicitly (DeLind, 2002; Lyson, 2004). Civic agriculture is defined by decision making that values and sustains local resources: by thinking of farming as the management of intricate ecological systems that include producers, consumer, labor, and land. Though in many ways an idealized notion of what agriculture should be, this train of thought is interesting because it makes clear that farming is (or can be) intricately linked to localities and places. In Lyson’s words: “Civic agriculture is a locally organized system of agriculture and food production characterized by networks of producers who are bound together by *place*. It embodies a commitment to developing and strengthening an economically, environmentally, and socially sustainable system of agriculture and food production that relies on local resources, and serves local markets and consumers. The imperative to earn a profit is filtered through a set of cooperative and mutually supporting social relations” (2004, p. 102). Civic agriculture also “takes up social, economic and geographic *spaces* not filled (or passed over) by industrial agriculture,” and “it relies more on indigenous, site-specific knowledge and less on a uniform set of best managements practices” (Lyson et al., 2008, p. 174, emphasis added). While different in their foci, both the literature on the agriculture of the middle and on civic agriculture draw attention to how different forms of production and ownership are linked to local communities in various ways. Both sets of contributions contrast the community linkages of these modes of production with those of industrial farming and animal rearing that rely on vast high-tech facilities, produce large amounts of waste, rely on underpaid and often illegal labor, and produce a homogenous product for distant processing facilities of commodity markets.

These debates on the community linkages of different types of farming are important here because of the chapter’s interest in the community linkages of policy regimes. Much more research is needed on the different ways in which policies affect farmers and communities; this chapter has only raised this question by outlining how different policy regimes had unequal effects across and within the United States and Russia. How economic pressures that characterize markets for agricultural commodities changed over the course of the twentieth century is well documented. Economies of scale at ever-larger operations became more important as markets for agricultural commodities expanded and became more homogenous (USDA Economic Research Service, 2013). At least to some extent, sustained state support for a particular type of farming operation, located in particular sites, is likely to have shaped the way these markets emerged. The challenge then is to learn about the community effects of farm policies and to revise policy regimes to take these lessons into account.

References

Abers, R. N., & Keck, M. E. (2013). *Practical authority: Agency and institutional change in Brazilian water politics.* New York, NY: Oxford University Press.

Allina-Pisano, J. (2007). *The Post-Soviet Potemkin village: Politics and property rights in the Black Earth.* New York, NY: Cambridge University Press.

Becker, A. (1987). *U.S.–Soviet Trade in the 1980s* (Rand Note No. N-2682-RC). The Rand Corporation. Retrieved from http://www.rand.org/content/dam/rand/pubs/notes/2009/N2682.pdf

Berk, G., Galvan, D. C., & Hattam, V. (Eds.). (2013). *Political creativity: Reconfiguring institutional order and change.* Philadelphia: University of Pennsylvania Press.

Deininger, K., & Byerlee, D. (2011). *The rise of large farms in land abundant countries* (Policy Research Paper No. 5588). Washington, DC: World Bank. Retrieved from http://elibrary.worldbank.org/doi/pdf/10.1596/1813-9450-5588

DeLind, L. B. (2002). Place, work, and civic agriculture: Common fields for cultivation. *Agriculture and Human Values, 19*(3), 217–224.

Environmental Working Group. (2012). EWG Farm Subsidy Database [Database]. Retrieved from http://farm.ewg.org/index.php

Espach, R. H. (2009). *Private environmental regimes in developing countries.* New York, NY: Palgrave Macmillan.

Fitzgerald, D. K. (2003). *Every farm a factory: The industrial ideal in American agriculture.* New Haven, CT: Yale University Press.

Goodman, D., DuPuis, E. M., & Goodman, M. K. (2012). *Alternative food networks: Knowledge, practice, and politics.* New York, NY: Routledge.

Gorbachev, M. S. (1996). *Memoirs.* New York, NY: Doubleday.

Hansen, J. M. (1991). *Gaining access: Congress and the farm lobby, 1919–1981.* Chicago, IL: University of Chicago Press.

Hedlund, S. (1989). *Private agriculture in the Soviet Union*. New York, NY: Routledge.

Ingersent, K. A., & Rayner, A. J. (1999). *Agricultural policy in Western Europe and the United States*. Northampton, MA: Edward Elgar Pub.

Ioffe, G., Nefedova, T., & Zaslavsky, I. (2006). *The end of peasantry? The disintegration of rural Russia*. Pittsburgh, PA: University of Pittsburgh Press.

Kirschenmann, F., Stevenson, G. W., Buttel, F., Lyson, T. A., & Duffy, M. (2008). Why worry about the agriculture of the middle? In T. A. Lyson, G. W. Stevenson, & R. Welsh (Eds.), *Food and the mid-level farm: Renewing an agriculture of the middle* (pp. 3–22). Cambridge, MA: MIT Press.

Lyson, T. A. (2004). *Civic agriculture: Reconnecting farm, food, and community*. Medford, MA: Tufts University Press.

Lyson, T. A., Stevenson, G. W., & Welsh, R. (2008). *Food and the mid-level farm: Renewing an agriculture of the middle*. Cambridge, MA: MIT Press.

MacDonald, J., Korb, P., & Hoppe, R. (2013). *Farm size and the organization of U.S. crop farming* (Economic Research Report No. ERR-152). USDA. Retrieved from http://www.ers.usda.gov/media/1156726/err152.pdf

Nikulin, A. (2003). Kuban kolkhoz between a holding and a hacienda. *Focaal*, *41*, 137–152.

Organisation for Economic Co-operation and Development (OECD). (2013). *Agricultural policy monitoring and evaluation 2013*. Paris: Organisation for Economic Co-operation and Development. Retrieved from http://www.oecd-ilibrary.org/content/book/agr_pol-2013-en

Overdevest, C. (2010). Comparing forest certification schemes: The case of ratcheting standards in the forest sector. *Socio-Economic Review*, *8*(1), 47–76.

Pallot, J., & Nefedova, T. (2007). *Russia's unknown agriculture: Household production in post-socialist rural Russia*. Oxford: Oxford University Press.

Rylko, D., Jolly, R., Khramov, I., & Uzun, V. (2008). Agroholdings: Russia's new agricultural operators. In Z. Lerman (Ed.), *Russia's agriculture in transition: Factor markets and constraints on growth* (pp. 95–133). Lanham, MD: Lexington Books.

Sabel, C. F., & Zeitlin, J. (2010). *Experimentalist governance in the European Union: Towards a new architecture*. New York, NY: Oxford University Press.

Sheingate, A. D. (2003). *The rise of the agricultural welfare state: Institutions and interest group power in the United States, France, and Japan*. Princeton, NJ: Princeton University Press.

Sumner, D. A. (2008). Agricultural Subsidy Programs. In *The Concise Encyclopedia of Economics* (2nd ed.). Library of Economics and Liberty.

Union of Concerned Scientists (UCS). (2012). *Ensuring the harvest: Crop insurance and credit for a healthy farm and food future.* Retrieved from http://www.ucsusa.org/food_and_agriculture/solutions/expand-healthy-food-access/ensuring-the-harvest.html

US Department of Agriculture (USDA). (2013). USDA ERS state export data. Retrieved from http://www.ers.usda.gov/data-products/state-export-data.aspx#.U1qrd8dwWfQ

US Department of Agriculture (USDA). (n.d.). Agricultural management assistance. Retrieved from http://www.nrcs.usda.gov/wps/portal/nrcs/main/national/programs/financial/ama/

US Department of Agriculture (USDA). Economic Research Service—AIS90. (n.d.). Retrieved from http://www.ers.usda.gov/publications/ais-agricultural-income-and-finance-outlook/ais90.aspx#.U3-poC-fufQ

Vassilieva, Y. (2012). *Russian federation: Agricultural development program 2013–2020* (Global Agricultural Information Network No. GAIN Report RS1270). USDA Foreign Agricultural Service. Retrieved from http://gain.fas.usda.gov/Recent%20GAIN%20Publications/Agriculture%20Development%20Program%202013-2020_Moscow_Russian%20Federation_11-6-2012.pdf

Wegren, S. K. (1998). *Agriculture and the state in Soviet and post-Soviet Russia.* Pittaburgh, PA: University of Pittsburgh Press.

Федеральная служба государственной статистики (Federal State Statistics Services). (2000, various years 2006). Основные показатли сельского хозяйства в России. Retrieved from http://www.gks.ru/wps/wcm/connect/rosstat_main/rosstat/ru/statistics/publications/catalog/doc_1140096652250

Chapter 8

Power and Place in Food Systems: From Global to Local

Savannah Larimore and Vaughn Schmutz

Introduction

At first glance, the globalization of food production, distribution, and consumption appears to bridge social and spatial distances, thereby providing greater access to food. In recent decades, consumers in affluent places such as the United States have seen a proliferation of ethnic dining options that make distant cuisines more accessible than ever before (Warde, 2000). Likewise, we are regularly reminded of how far many foods in a complex, global food system must travel before reaching our plates (Schnell, 2013). Americans might enjoy a fruit salad, for instance, that features pineapple from the Philippines, cantaloupe from Guatemala, and grapes from Chile. Yet the wealth and diversity of food offerings that our global food system affords in some places is countered by a lack of plenty in others. Rather than creating abundance and variety for all food consumers, the contemporary agri-food industry perpetuates inequalities in food access, reproduces food insecurity, homogenizes the global food supply, and poses risks to the natural environment. In attempting to eradicate the constraints of place, the food systems on which we rely too often exacerbate or generate stark place-based inequalities.

In this chapter, we highlight examples of these inequalities, showing how local spaces are linked to the power dynamics of the global food system. Our goal is to demonstrate how place matters—at the global, national, and local levels—as well as how food-related inequalities linked to place intersect with social characteristics such as race, class, and gender. When we speak of place, we are not only referring to a specific geographic location, but to the unique history and characteristics that give a locale meaning and value (Gieryn, 2000).

Additionally, while a variety of alternative food movements, such as those advocating "local" products, aim to counteract the negative consequences of our global agri-food system, we argue that such efforts too often overlook issues of inequality that are associated with place. We conclude by offering some place-based suggestions for moving toward food systems that are more socially and environmentally just, equitable, and sustainable.

Global Place-Based Inequalities

In the view of many scholars and concerned observers, the global agri-food system is in a state of crisis on many fronts. In the years leading up to the 2008 global food crisis, a combination of rising oil prices, biofuel demands, and trade shocks created volatility in the international food market, resulting in substantial price increases for staple crops such as rice and wheat (Headey & Fan, 2010; United Nations (UN), 2011). Developing nations that rely heavily on imports are most susceptible to fluctuations in food prices as even modest price increases pose serious obstacles to food security for low-income citizens in the global South, who must spend a larger share of their meager earnings on staple crops, with little money left over for other necessities (FAO, 2013; Marktanner & Noiset, 2013; Mazzeo, 2009). The most recent report from the Food and Agriculture Organization (FAO) of the United Nations (UN) indicates that between 2011 and 2013, roughly 842 million—about one out of every eight people in the world—suffered from chronic hunger, the vast majority of whom reside in developing countries (FAO, 2013). Although this represents a slight improvement since the height of the 2008 global food crisis, it reveals ongoing inequalities in food access. Thus, place matters at the broadest level as food-related inequalities are readily observable between developing and developed nations.

Within developing countries, spatial and social disparities place certain communities at heightened risk for food insecurity and hunger. Whereas rural communities have traditionally been able to achieve greater levels of security amid poverty by supplementing their food purchases with subsistence agriculture, these practices have become less common over time as rural farms shrink or disappear and younger generations migrate from rural communities to more densely populated areas in search of wage labor (Fazzino & Loring, 2009; Mazzeo, 2009). While rates of food insecurity have traditionally been higher in rural areas (Ruel et al., 1998), researchers and policy makers are beginning to shift their focus toward urban areas as a result of recent population shifts. Developing nations are experiencing rapid rates of urbanization that heighten spatial imbalance and urban poverty (Smith, Obeid, & Jensen, 2000). During the 2008 crisis, the vulnerability of poor, urban residents was especially apparent. These groups are largely food buyers rather than food producers and consistently suffer most from unpredictable food prices (Zezza & Tasciotti, 2010).

In both the rural and the urban areas of developing countries, women and children have particularly high rates of malnourishment and food insecurity. In patriarchal societies, women face considerable obstacles to procuring an adequate quantity and quality of food to feed their family, given social restrictions (Githinji, 2009; Levay, Mumtaz, Rashid, & Willows, 2013). In Buhaya (northwestern Tanzania), for example, women are primarily responsible for farmwork and food provision, yet they lack equal representation in family and community decision making as well as access to resources such as money, land, and education, which ultimately places them at greater risk of hunger during food shortages (Githinji, 2009). In Bangladesh, poor women experience particularly high rates of maternal malnutrition. Rising food prices amid recent global crises compounded this problem due to women's limited involvement in family decision making, lack of access to family food budgets, and food allocation practices that privilege male household members (Levay et al., 2013). Likewise, young children and adolescents experience greater instances of food insecurity or hunger, which corresponds with poorer health and can have cumulative effects over the life course (FAO, 2013; Hadley, Belachew, Lindstrom, & Tessema, 2009). The circumstances of food environments in the global South perpetuate a cycle of poverty, hunger, and illness that disproportionately impacts certain groups.

The global place-based inequalities described above have been compounded by economic and cultural developments around the globe. After the 2008 crisis, neoliberal restructuring of food production in developing countries culminated in economic policies that favored open markets, allowing for foreign direct investment and free trade while limiting state regulation of the private sector. Whereas social scientific theories offer competing explanations for the rise and spread of neoliberal economic reforms around the world (see Chorev, 2010) and highlight the divergent paths by which developing countries have embraced such policies (Fourcade-Gourinchas & Babb, 2002), it is clear that these changes have greatly enhanced global economic integration with profound consequences for the production, distribution, and consumption of food around the world. As international institutions—such as the World Trade Organization (WTO) or the International Monetary Fund (IMF)—promote such policies to governments, and as elites in developing countries encounter neoliberal ideas through higher education (Schofer & Meyer, 2005), integration into the global economy gains wider appeal.

However, some argue that the neoliberal policies promoted by treaties such as the WTO Agreement on Agriculture have made food security more tenuous for the most vulnerable populations in the global South (Gonzalez, 2002). This occurs partly because open markets and free trade shift the balance of power away from small producers (who are unable to compete in an international market) and developing countries (who are prevented from protecting their domestic agriculture)—while the power of industrialized countries and transnational corporations that operate in the global agri-food system grows (Martinez-Gomez, Aboites-Manrique, & Constance, 2013). Thus, in many places, the global food system has contributed to the decline and marginalization of small farmers, a shift in agricultural ownership to transnational conglomerates, and heightened rural poverty and inequality.

To illustrate the impacts of such policies, we draw from the vast literature on the North American Free Trade Agreement (NAFTA) and subsequent changes to production in Mexico, focusing specifically on corn (Avalos & Graillet, 2013; Keleman, 2009), poultry (Martinez-Gomez et al., 2013), and agave production for tequila (Bowen & Gaytán, 2012). Post-NAFTA, neoliberal reform in favor of free agricultural trade had the effect of reducing the price of corn and increasing the price of tortillas, which hurt both Mexican farmers and consumers (Avalos & Graillet, 2013). Small farms were often unable to compete on the international market, and over one million farmers are estimated to have migrated to urban areas or the United States in the first decade after NAFTA took effect. As a result, corn production became more concentrated in the hands of large, industrial producers. In addition to heightening rural food insecurity and poverty in Mexico, this also reduced the genetic diversity of the corn as small farmers had traditionally acted as stewards of maize varieties (Keleman, 2009).

In the case of poultry, Mexico has increasingly adopted the US model of poultry production under the influence of the IMF and NAFTA, despite criticisms of the modern poultry industry (Martinez-Gomez et al., 2013). As with corn, this has led to the consolidation of the poultry industry, as small-scale and many medium-sized companies went out of business while large-scale producers like Bachoco and Tyson steadily increased their market share. While this has benefited transnational corporations (TNCs) that operate in the poultry industry, it has further marginalized rural Mexicans. Agave production for Mexican tequila has followed a similar trajectory with TNCs increasingly controlling and profiting from a cultural tradition that had been preserved by small farmers and distilleries for many generations (Bowen & Gaytán, 2012). In this case, TNCs capitalized on the state's interest in promoting a product (i.e., tequila) that had been linked to the cultural heritage and national identity of Mexico, but this came at the expense of small producers who possessed greater knowledge of agave production. In these and other examples, the evolution of the global agri-food system has simultaneously minimized the constraints of place for developed countries and increased or created new place-specific inequalities for nations and individuals in the global South.

While economic restructuring exposes and exacerbates food-related inequalities around the world, cultural changes also have consequences for global food security. At the beginning of the chapter, we alluded to the expansive food options available to privileged consumers in some places but also described how the diversity of crops is threatened by the growing consolidation of food production. Further reinforcing such trends is the growing similarity of food consumption practices around the globe. As the global agri-food system expands, Western dietary preferences are increasingly adopted throughout the world. Though this has tended to diversify the diet within countries in recent decades, it has contributed to an overall trend of diminishing diversity in the global food supply (Khoury et al., 2014). In other words, the variety available in the global food basket has declined over the past half century. Aside from the environmental and health consequences of this trend that is narrowing the food supply, it raises concerns about food security as the world population becomes increasingly reliant on a smaller number of global crop commodities. As seen above, vulnerable populations suffer the most when shocks to global food markets lead to price volatility in staple crops, which suggests that such groups are at greater risk as our food supply becomes less varied.

Domestic Place-Based Inequalities

The transformative processes described above have positioned the United States favorably as one of the world's principal food producers (Cooper, 1986). However, these developments have also led to disproportionate levels of food insecurity and hunger across certain sections of the American population that mirror the prevalence of hunger and malnourishment in the developing world (Holt-Gimenez & Wang, 2011). While much research has been dedicated to understanding how the global food regime negatively impacts the global South, it is also necessary to consider the relationship between the agricultural sector in the United States and global food markets as well as the impact of this relationship on communities and individuals. Within the United States, place-based inequalities exist between rural and urban communities as well as within these communities, as social characteristics, such as race, class, and gender compound the inequalities of place and produce profound disparities in levels of food insecurity (Coleman-Jensen, Nord, & Singh, 2013).

In the twentieth century, the American agricultural sector underwent major restructuring that resulted in the expansion of large-scale mechanized farming operations specializing in commodity crops. The increase in large-scale farms coincided with a decrease in the overall number of farms as many smaller farms disappeared. Coupled with technological advances, the percent of the US labor force employed in the agricultural sector dwindled from 41 percent in the early 1900s to less than 2 percent at the turn of the twenty-first century (Dimitri, Effland, & Conklin, 2005). As a result of changes to the US agricultural sector, rural communities also experienced economic and social transformations.

The drastic decrease in farm-related employment opportunities coincided with a decrease in the rural population nationwide (Dimitri et al., 2005). Individuals and families with enough human capital and other resources migrated from rural areas to urban centers to seek employment (Berardi, 1981; Lobao & Meyer, 2001), leaving many poor and undereducated rural residents behind to face unemployment, lower wages, increased income inequality, and other negative consequences (Albrecht, Albrecht, & Albrecht, 2000; Crowley & Roscigno, 2004; Gouveia, 1994; Tolbert & Lyson, 1992). While the direct negative effects of large-scale agriculture on rural communities have been the subject of considerable debate (Barnes & Blevins, 1992; Peters, 2002; Durrenberger & Thu, 1996; Green, 1985; Lobao, Schulman, & Swanson, 1993), it is clear that the landscape of rural America has been forever changed.

Economic restructuring in rural America has produced one of the greatest ironies in our food system: rural areas, where much of the world's food is produced, experience rates of food insecurity and hunger that are above the US national average, with the rural South being the most vulnerable (Coleman-Jensen et al., 2013; Stuff et al., 2004). As previously noted, when agricultural jobs disappeared, so did many middle- to upper-income rural residents, taking with them their economic resources and patronage of local businesses (Albrecht, Albrecht, & Murguia, 2005). Without a reliable customer base, many small food retailers went out of business, further depleting the food environment in rural communities. The combination of these factors has led to increased poverty in rural communities (Tickamyer & Duncan, 1990) and produced food deserts (Morton & Blanchard, 2007), meaning that the area is not only low-income but that many residents live at least ten miles away from a full-service grocery store (Dutko, Ploeg, & Farrigan, 2012). This distance is especially daunting for low-income consumers who are without access to reliable transportation—public or private (Bletzacker, Holben, & Holcomb, 2009; Connell et al., 2007). This is not to say that food retailers do not exist in rural areas. However, more often than not, the few stores that service rural communities are small and do not offer a wide variety of fresh, nutritious, or affordable foods (Hosler, 2009; Liese, Weis, Pluto, Smith, & Lawson, 2007).

Although the constraints of place are highly pronounced in rural areas, food insecurity is not a distinctly rural problem. Researchers have extensively documented the existence of food deserts in low-income urban areas where many residents must travel well outside their neighborhoods to reach a full-service supermarket (Dutko et al., 2012). What differentiates urban food environments from rural ones is that urban areas often have an abundance of small food retailers (Powell, Slater, Mirtcheva, Bao, & Chaloupka, 2007). At first glance, it would appear that urban residents have more than enough access to food. However, closer inspection of these small food retailers depicts what several scholars refer to as a "food mirage," in which urban food retailers are plentiful but their shelves are stocked with calorie-dense, nutritionally deficient, poor quality, expensive foods (Breyer & Voss-Andreae, 2013; Hosler, Rajulu, Ronsani, & Fredrick, 2008; Short, Guthman, & Raskin, 2007). It is in the instance of the urban food mirage that we see the effects of the global food regime in urban centers across the United States.

As a result of changes in US agricultural policies in the 1970s, commodity crops—namely, corn, grain, and soy—have become overly abundant, cheap to buy, and cheap to process (Pollan, 2003). While this transformation has certainly had global repercussions (Hawkes, 2006), it has changed the domestic food environment in the United States as well, creating an abundance of cheap, processed, calorie-rich, and nutrient-poor food products (Popkin, Duffey, & Gordon-Larsen, 2005). Additionally, food-price data shows a substantial increase in the price of healthy foods, such as vegetables, fruits, and fish, relative to increases in the price of unhealthy, processed foods (Gelbach, Klick, & Stratman, 2007). As with rural shoppers, if urban residents want to stock their pantries with nutritious and affordable foods, they must travel well outside their immediate areas and commit more time, money, and effort to large shopping trips than their affluent, suburban, and typically white counterparts (Alwitt & Donley, 1997; Block, Chavez, & Birgen, 2008; Bodor, Rice, Farley, Swalm, & Rose, 2010; Munoz-Plaza, Filomena, & Morland, 2007; Yeh et al., 2008).

Both in urban and in rural communities, degrees of food security vary by social location in terms of class, race, gender, and other social characteristics. Income and food security are positively correlated, as an increase in income provides access to resources such as transportation that negate the distance to supermarkets or allow wealthier consumers to purchase sufficient amounts of food (Coleman-Jensen et al., 2013; Cotterill & Franklin, 1995). Whereas individual income plays a crucial role in food security, the economic composition of one's neighborhood also dictates access to certain foods. Residents in low-income neighborhoods, regardless of their own income, have fewer and smaller food retailers in the immediate area, fewer healthy options within local stores, and may pay more when shopping locally for similar items available at faraway supermarkets (Alwitt & Donley, 1997; Chung & Myers, 1999; Zenk et al., 2005).

Further complicating the inequalities of place, the historical impacts of residential racial segregation in the United States intertwine with the food environment such that predominantly minority communities have less access to food retailers that provide fresh, healthy, and affordable foods (Block & Kouba, 2006; Raja, Ma, & Yadav, 2008). Whereas neighborhood racial composition is certainly not a proxy for neighborhood economic composition, predominantly minority communities face many of the same obstacles as low-income communities. In comparison to racially mixed or predominantly white neighborhoods, African American communities have roughly half as many full-service supermarkets available in the immediate area (Powell et al., 2007). Although the availability of supermarkets is slightly better for predominantly Hispanic neighborhoods, persons residing in predominantly minority communities must travel further to reach full-service supermarkets located in suburban, predominantly white communities (Zenk et al., 2005). The food retailers that are available in minority communities are often convenience stores or small grocery stores that have fewer healthy, fresh, whole-food options (Block & Kouba, 2006; Franco, Diez Roux, Glass, Caballero, & Brancati, 2008). In addition, when healthy options or whole foods are available at small retailers, they tend to be more expensive than similar items sold at full-service supermarkets (Chung & Myers, 1999). Living in a predominantly minority community often means traveling further for adequate foods or paying more for poorer quality foods.

As was the case in developing nations, gender also plays a crucial role in food access as women—especially in urban communities—report unique barriers to food access as a result of neighborhood disorder. Women, who tend to be the primary food shoppers in households, report feeling unsafe in neighborhood stores that are poorly monitored and unkempt (Bader, Purciel, Yousefzadeh, & Neckerman, 2010; Block et al., 2008; Zenk et al., 2011). In the first author's own research, from interviewing food-stamp recipients in an urban food desert, several female shoppers reported a need to "protect their purse" while shopping at food retailers in their local communities (Larimore, 2014). Female-headed households with children are especially vulnerable to elevated rates of food insecurity (Coleman-Jensen et al., 2013) as women maintain gendered expectations to be responsible for grocery shopping, preparing meals, and feeding others in their households (DeVault, 1994). Mothers in food-insecure households have been shown to adopt a number of food-provisioning strategies, some of which—including skipping their own meals or eating less to give their children enough food—are detrimental to their own health (McIntyre et al., 2003). Although income, race, and gender are not the only social characteristics that further complicate barriers to food access in the United States, it is clear that the constraints of place are more salient for some than for others.

Place-Based Alternatives

Given that so many inequalities in our global and domestic food system are associated with place and compounded by social location (e.g., race, class, gender), it seems reasonable to look for solutions to these challenges from a place-based perspective as well. In recent decades, there has been a growing push in many developed countries among grassroots organizations, food producers, consumers, and scholars that advocates relocalizing the food system (Rogers & Fraszczak, 2014; Schnell, 2013). In its ideal form, the local food movement intends to create economically and environmentally sustainable food economies that enhance community engagement and promote social equity in a given place (Feenstra, 2002). In many ways, the local food movement can claim some success as it has gained widening popularity among consumers, which creates markets for small-scale local farmers and connects those farmers more directly to the customers they serve (Allen & Wilson, 2008). Yet as Schnell (2013) argues, "local" food advocates can too often be narrowly aimed at reducing food miles (i.e., the distance food travels before it reaches our plates) and lose the place-based ideals that add deeper significance to eating local. Furthermore, many organizations that are generally celebrated by the local food movement, such as farmers markets or community supported agriculture (CSA) groups, overlook the deep-rooted inequalities within the dominant food system outlined in this chapter.

It is clear that markets have been quick to respond to the demands of consumers who have embraced the alternative food movement with new products, making "local" one among any number of commodity characteristics (Weber, Hienz, & DeSoucey, 2008). However, it is this very rise in popularity that has left the alternative food movement open to the same neoliberal market processes that prevail in the dominant global food system. Far too much emphasis has been placed solely on economic relationships and market exchanges, which reinforces inequalities by privileging consumers with resources and fails to invigorate local communities with a sense of place (DeLind, 2011). Arguably, this is especially the case in farmers markets, CSAs, and even large chain grocery stores that sell local food items where customers are able to "vote with their dollar" (Johnston, 2008). Not only is this transformation contradictory to the aims and values of the alternative food movement, but also these market-based solutions often remain disembedded from local spaces, favor affluent consumers, and make limited attempts to incorporate marginalized consumers into the fold (Hinrichs & Allen, 2008). Thus, despite its achievements and promise, many local food initiatives currently fall short of the holistic, regenerative (i.e., self-sustaining over time) food systems that were initially envisioned by earlier generations of scholars and activists (Dahlberg, 1993).

In calling attention to the shortcomings of the local food movement, scholars and other critics highlight a need for alternative markets to actualize a sense of "embeddedness" by which social relations uphold economic institutions (Granovetter, 1985). Among other things, embeddedness refers to the degree to which informal relationships between persons in a given place shape the everyday functions of an economic structure, such as an alternative food market.

In the case of farmers markets, this often manifests itself in an idealized concept of "community" by which farmers, market organizers, customers, and community members share equally in sustaining the market, not only economically but also socially (Hinrichs, 2000). Although this shared responsibility aligns with the goals of relocalization efforts, many alternative markets have struggled to make embeddedness a reality.

Much of this can be attributed to the aforementioned reproduction of neoliberal market processes in alternative food systems. By reducing a sense of community and place—based in shared meaning and history—to an economic exchange between producer and consumer, local food becomes less of a political statement against the dominant agricultural model and more of a niche consumer market (DeLind, 2011). In this way, alternative food consumers are open to the same critiques as conventional food consumers, who are perceived as unaware of or disinterested in the complex agricultural and economic processes producing the food on their dinner plates. While establishing direct food markets is a step in the appropriate direction, creating food citizens—persons who acknowledge the rights and responsibilities of living and eating in a given place—will require more community-based approaches. Indeed, many have suggested that alternative food systems can become more regenerative and inclusive by aiming to empower food citizens who support democratic, equitable, and socially sustainable practices rather than simply seeking to serve food consumers (Wilkins, 2005).

However, the collective voice of the local food movement rarely represents the voice and values of all community members. The local food movement is overwhelmingly white, college-educated, upper-income, and liberal, which can make it more difficult to engage low-income and minority communities (Alkon & McCullen, 2011; Brown, Dury, & Holdsworth, 2009; Macias, 2008). By way of the market processes that privilege certain consumers over others, many alternative food markets embody what Alkon and McCullen (2011) refer to as a "liberal, affluent, white culture." Put simply, spaces such as farmers markets that promote local food often privilege affluent customers economically and culturally, thereby reinforcing many of the racial and class inequalities evident in the conventional food system. Failing to acknowledge the sustainable practices of low-income consumers implies that there is only one "correct" way to do alternative consumption (Johnston, Szabo, & Rodney, 2011). This can be especially problematic when movement organizers attempt to relocalize the food environment in food deserts and underserved communities.

Aside from economic constraints to equal participation by minorities and working-class consumers in alternative food systems, recent studies have found that producers and consumers within these systems actively create cultural barriers as well (Beagan & Chapman, 2012; Fotopoulos & Krystallis, 2002; Guthman, 2008b). As the popularity of alternative food systems has increased, a "dominant ethical eating repertoire" has developed, which advocates patterns of consumption that align with the ethics and practices of alternative food systems. Given the demographics of the typical alternative-food-system participant, this repertoire embodies a culture that is affluent, liberal, educated, and white. Additionally, knowledge of alternative food systems is activated as a form of cultural capital, labeling the consumer as ethical and indicating privilege (Johnston et al., 2011). Lacking this capital, minority and low-income consumers report discomfort in alternative-food-system spaces such as farmers markets, where they actively label the foods available in these spaces as "not for them" (Alkon & McCullen, 2011; Guthman, 2008b).

To illustrate this point, we draw on an example from the first author's (Larimore, 2014) field research at two urban farmers markets in North Carolina. One market—which operated next to a food desert and accepted food stamps to reduce spatial and economic barriers to participation—failed to produce a customer base that was representative of the residents in the immediate area. In interviews with market organizers, many felt a need to educate low-income and minority residents on how to eat sustainably and healthfully—a discourse that seems to permeate the alternative food movement (Guthman, 2008b; Guthman, 2008c; Lyson, 2014). However, market organizers had made minimal efforts to seek out the opinions and desires of marginalized consumers; often, potential social barriers to participation were disregarded and consumer behavior was reduced to individual choice (Larimore, 2014). In this way, the market reaffirmed the neoliberal concept of "choice" (Allen & Wilson, 2008; Guthman, 2008a) and failed to produce a horizontally embedded social structure where the considerations of all supply-chain actors were validated rather than just those of customers with resources or farmers (Bowen, 2011; Hinrichs, 2000). By contrast, a farmers market in a similar neighborhood in the same city was successful in achieving broader participation among local residents, including food-stamp recipients, and maintained a closer connection to the surrounding community. In addition to the social embeddedness that characterizes this market, customers felt that their cultural conceptions of alternative food systems were valued in this setting. Further research is needed to disentangle the conditions under which social and cultural barriers to participation are more likely to be bridged in these ways, but it is clear that participatory communities are vital to achieving more equitable outcomes in food access and security.

Conclusions and Observations

To be clear, we do not wish to dismiss the remarkable successes of relocalization efforts in alternative food movements. However, we would argue that it is essential to acknowledge that even place-based solutions to global and domestic food issues can reproduce existing inequalities and can potentially produce new disparities in access to food. Above, we have alluded to the need to overcome a variety of economic, social, and cultural barriers to achieve greater participation that embeds food citizens in their local communities. We conclude with some additional suggestions for policy makers and market organizers that would move us closer to such ambitious goals. First, as Trubek and Bowen (2008) have suggested, federal- or state-level regulations need to be implemented that protect the quality of place-based products as well as the livelihoods of the communities that produce them. In doing so, not only is the economic viability of a place-specific product secured, but also the economic livelihoods of the farmers who produce these products and the communities they live in. Such products preserve place in a broad sense as they are often tied to the cultural heritage and identity of the local community. Although government policy can undermine local producers by favoring TNCs (e.g., the case of agave production in Mexico, discussed above) it can also help preserve place-based designations in the face of potentially destabilizing forces (Bowen, 2011; DeSoucey, 2010).

In addition, market organizers can make a concerted effort to value social inclusion above economic viability. To do so, marginalized consumers must be included not only in the customer base, but also in broader conversations about market processes and community engagement. This can be achieved in a variety of ways: including customers in decisions about hours of operation; creating communal spaces for community members to interact, even if they do not purchase any food items; and listening to complaints or concerns of community members. Although this represents a significant shift away from focusing on conceptions of community that are rooted in free choice and market exchange, there may be reason for organizers and producers to be optimistic about eaters' desire to connect with place. Whereas popular media may focus on relatively narrow, distance-based depictions of local eating, Schnell (2013) found that CSA participants expressed a broad range of social, environmental, and economic motivations for their involvement. For CSA participants, place was central to creating embedded relationships and to their member experience. By shifting the focus from simply eating local to experiencing place, there is potential to move closer to food systems that are more socially and environmentally just, equitable, and sustainable.

References

Ahern, M., Brown C., &Dukas, S. (2011). A national study of the association between food environments and county-level health outcomes. *Journal of Rural Health*, *27*(4), 367–379.

Albrecht, D. E., Albrecht, C. M., & Albrecht, S. L. (2000). Poverty in nonmetropolitan America: Impacts of industrial, employment, and family structure variables. *Rural Sociology*, *65*(1), 87–103.

Albrecht, D. E., Albrecht C. M., & Murguia, E. (2005). Minority concentration, disadvantage, and inequality in the nonmetropolitan United States. *Sociological Quarterly*, *46*(3), 503–523.

Alkon, A. H. (2012). *Black, white, and green: Farmers markets, race, and the green economy*. Athens: University of Georgia Press.

Alkon, A. H., & McCullen, C. G. (2011). Whiteness and farmers markets: Performances, perpetuations... contestations? *Antipode*, *43*(4), 937–959.

Allen, P., & Wilson, A. B. (2008). Agrifood inequalities: Globalization and localization. *Development*, *51*, 534–540.

Alwitt, L. F., & Donley, T. D. (1997). Retail stores in poor urban neighborhoods. *Journal of Consumer Affairs*, *31*(1), 139–164.

Avalos, A., & Graillet, E. (2013). Corn and Mexican agriculture: What went wrong? *American Journal of Economics and Sociology*, *72*(1), 145–178.

Bader, M. D., Purciel, M., Yousefzadeh, P., & Neckerman, K. M. (2010). Disparities in neighborhood food environments: Implications of measurement strategies. *Economic Geography*, *86*(4), 409–430.

Barnes, D., & Blevins, A. (1992). Farm structure and the economic well-being of nonmetropolitan counties. *Rural Sociology*, *57*(3), 333–346.

Beagan, B. L., & Chapman, G. E. (2012). Meanings of food, eating and health among African Nova Scotians: 'Certain things aren't meant for Black folk.' *Ethnicity & Health*, *17*(5), 513–529.

Berardi, G. M. (1981). Socio-economic consequences of agricultural mechanization in the United States: Needed redirections for mechanization research. *Rural Sociology*, *46*, 483–504.

Bletzacker, K. M., Holben, D. H., & Holcomb, J. P., Jr. (2009). Poverty and proximity to food assistance programs are inversely related to community food security in an Appalachian Ohio region. *Journal of Hunger & Environmental Nutrition*, *4*(2), 172–184.

Block, D., & Kouba, J. (2006). A comparison of the availability and affordability of a market basket in two communities in the Chicago area. *Public Health Nutrition, 9*(7), 837–845.

Block, D., Chavez, N., & Birgen, J. (2008). Finding food in Chicago and the suburbs: The report of the northeastern Illinois community food security assessment. *Chicago State University Frederick Blum Neighborhood Assistance Center and University of Illinois-Chicago School of Public Health, Division of Community Health Sciences.* Retrieved from http://www.csu.edu/nac/documents/FindingFoodinChicagoandtheSuburbsReporttothePublic2008.pdf

Block, D. R., Chávez, N., Allen, E., & Ramirez, D. (2012). Food sovereignty, urban food access, and food activism: Contemplating the connections through examples from Chicago. *Agriculture and Human Values, 29*(2), 203–215.

Bodor, J. N., Rice, J. C., Farley, T. A., Swalm, C. M., & Rose, D. (2010). Disparities in food access: Does aggregate availability of key foods from other stores offset the relative lack of supermarkets in African-American neighborhoods? *Preventive Medicine, 51*(1), 63–67.

Bowen, S. (2011). The importance of place: Re-territorialising embeddedness. *Sociologia Ruralis, 51*(4), 325–348.

Bowen, S., & Gaytán, M. S. (2012). The paradox of protection: National identity, global commodity chains, and the tequila industry. *Social Problems, 59*(1), 70–93.

Bowen, S., & Zapata, A. V. (2009). Geographical indications, *terroir*, and socioeconomic and ecological sustainability: The case of tequila. *Journal of Rural Studies, 25*(1), 108–119.

Breyer, B., & Voss-Andreae, A. (2013). Food mirages: Geographic and economic barriers to healthful food access in Portland, Oregon. *Health & Place, 24*, 131–139.

Brown, E., Dury, S., & Holdsworth, M. (2009). Motivations of consumers that use local, organic fruit and vegetable box schemes in Central England and Southern France. *Appetite, 53*(2), 183–188.

Chorev, N. (2010). On the origins of neoliberalism: Political shifts and analytical challenges. In K. T. Leicht and J. C. Jenkins (Eds.), *Handbook of politics: State and society in global perspective* (pp. 127–144). New York, NY: Springer.

Chung, C., & Myers, S. L. (1999). Do the poor pay more for food? An analysis of grocery store availability and food price disparities. *Journal of Consumer Affairs, 33*(2), 276–296.

Coleman-Jensen, A., Nord, M., & Singh, A. (2013). *Household food security in the United States in 2012* (ERS Publication No. ERR-155). Washington, DC: US Department of Agriculture, Economic Research Service. Retrieved from http://www.ers.usda.gov/publications/err-economic-research-report/err155.aspx

Connell, C. L., Yadrick, M. K., Simpson, P., Gossett, J., McGee, B. B., & Bogle, M. L. (2007). Food supply adequacy in the Lower Mississippi delta. *Journal of Nutrition Education and Behavior, 39*(2), 77–83.

Cooper, C. F. (1986). American agriculture and the world community. In K. Dahlberg (Ed.), *New directions for agriculture and agricultural research: Neglected dimensions and emerging alternatives* (pp. 65–80). Totowa, NJ: Rowman & Allanheld.

Cotterill, R. W., & Franklin, A. W. (1995). *The urban grocery store gap.* (Issue Paper No. 8). Storrs, CT: Food Marking Policy Center. Retrieved from http://www.fmpc.uconn.edu/publications/ip/ip8.pdf

Crowley, M., & Roscigno, V. J. (2004). Farm concentration, political-economic process, and stratification: The case of the North Central U.S. *Journal of Political & Military Sociology, 32*(1), 133–155.

Dahlberg, K. A. (1993). Regenerative food systems: Broadening the scope and agenda of sustainability. In P. Allen (Ed.), *Food for the future: Conditions and contradictions of sustainability* (pp. 75–102). New York, NY: John Wiley and Sons.

DeLind, L. B. (2011). Are local food and the local food movement taking us where we want to go? Or are we hitching our wagons to the wrong stars? *Agriculture and Human Values, 28*(2), 273–283.

DeSoucey, M. (2010). Gastronationalism: Food traditions and authenticity politics in the European Union. *American Sociological Review, 75*(3), 432–455.

DeVault, M. L. (1994). *Feeding the family: The social organization of caring as gendered work.* Chicago, IL: University of Chicago Press.

Dimitri, C., Effland, A., & Conklin, N. (2005). *The 20th century transformation of US agriculture and farm policy* (ERS Publication No. EIB-3). Washington, DC: U.S. Department of Agriculture, Economic Research Service. Retrieved from http://www.ers.usda.gov/media/259572/eib3_1_.pdf

Durrenberger, E., & Thu, K. (1996). The expansion of large scale hog farming in Iowa: The applicability of Goldschmidt's findings fifty years later. *Human Organization, 55*(4), 409–415.

Dutko, P., Ver Ploeg, M., & Farrigan, T. (2012). *Characteristics and influential factors of food deserts* (ERS Report No. ERR-140). Washington, DC: U.S. Department of Agriculture, Economic Research Service. Retrieved from http://www.ers.usda.gov/publications/err-economic-research-report/err140.aspx

Food and Agriculture Organization (FAO). (2013). *The state of food insecurity in the world 2013: The multiple dimensions of food security.* Rome, FAO. Retrieved from http://www.fao.org/docrep/018/i3434e/i3434e00.htm

Fazzino, D. V., & Loring, P. A. (2009). From crisis to cumulative effects: Food security challenges in Alaska. *NAPA Bulletin, 32*(1), 152–177.

Feenstra, G. (2002). Creating space for sustainable food systems: Lessons from the field. *Agriculture and Human Values, 19*(2), 99–106.

Fotopoulos, C., & Krystallis, A. (2002). Organic product avoidance: Reasons for rejection and potential buyers' identification in a countrywide survey. *British Food Journal, 104*(3/4/5), 233–260.

Fourcade-Gourinchas, M., & Babb, S. L. (2002). The rebirth of the liberal creed: Neoliberalism in four countries. *American Journal of Sociology, 108*(3), 533–579.

Franco, M., Diez Roux, A. V., Glass, T. A., Caballero, B., & Brancati, F. L. (2008). Neighborhood characteristics and availability of healthy foods in Baltimore. *American Journal of Preventive Medicine, 35*(6), 561–567.

Gelbach, J. B., Klick, J., & Stratmann, T. (2007). Cheap donuts and expensive broccoli: The effect of relative prices on obesity. *Social Science Research Network*. Retrieved from http://papers.ssrn.com/sol3/papers.cfm?abstract_id=976484

Gieryn, T. F. (2000). A space for place in sociology. *Annual Review of Sociology, 26*, 463–496.

Githinji, V. (2009). Food insecurity in Buhaya: The cycle of women's marginalization and the spread of poverty, hunger, and disease. *NAPA Bulletin, 32*(1), 92–114.

Gonzalez, C. G. (2002). Institutionalizing inequality: The WTO agreement on agriculture, food security, and developing countries. *Columbia Journal of Environmental Law, 27*(2), 433–490.

Gordon, C., Purciel-Hill, M., Ghai, N. R., Kaufman, L., Graham, R., & Van Wye, G. (2011). Measuring food deserts in New York City's low-income neighborhoods. *Health and Place, 17*(2), 696–700.

Gouveia, L. (1994). Global strategies and local linkages: The case of the U.S. meatpacking industry. In Alessandro Bonanno (Ed.), *From Columbus to ConAgra: The globalization of agriculture and food* (pp. 125–148). Lawrence: University Press of Kansas.

Granovetter, M. (1985). Economic action and social structure: The problem of embeddedness. *American Journal of Sociology, 91*(3), 481–510.

Green, G. P. (1985). Large-scale farming and the quality of life in rural communities: Further specification of the Goldschmidt hypothesis. *Rural Sociology, 50*(2), 262–274.

Guthman, J. (2008a). Neoliberalism and the making of food politics in California. *Geoforum, 39*(3), 1171–1183.

Guthman, J. (2008b). "If they only knew": Color blindness and universalism in California alternative food institutions. *Professional Geographer, 60*(3), 387–397.

Guthman, J. (2008c). Bringing good food to others: Investigating the subjects of alternative food practice. *Cultural Geographies, 15*(4), 431–447.

Hadley, C., Belachew, T., Lindstrom, D., & Tessema, F. (2009). The forgotten population? Youth, food insecurity, and rising prices: Implications for the global food crisis. *NAPA Bulletin, 32*(1), 77–91.

Hawkes, C. (2006). Uneven dietary development: Linking the policies and processes of globalization with the nutrition transition, obesity and diet-related chronic diseases. *Globalization and Health, 2*, 4.

Headey, D., & Fan, S. (2010). *Reflections on the global food crisis. How did it happen? How has it hurt? And how can we prevent the next one?* (No. 165).

International Food Policy Research Institute. Retrieved from http://www.ifpri.org/sites/default/files/publications/rr165.pdf

Hinrichs, C. C. (2000). Embeddedness and local food systems: Notes on two types of direct agricultural market. *Journal of Rural Studies, 16*(3), 295–303.

Hinrichs, C. C., & Allen, P. (2008). Selective patronage and social justice: Local food consumer campaigns in historical context. *Journal of Agricultural and Environmental Ethics, 21*(4), 329–352.

Holt-Giménez, E., & Wang, Y. (2011). Reform or transformation? The pivotal role of food justice in the U.S. food movement. *Race/Ethnicity: Multidisciplinary Global Contexts, 5*(1), 83–102.

Hosler, A. S. (2009). Retail food availability, obesity, and cigarette smoking in rural communities. *Journal of Rural Health, 25*(2), 203–210.

Hosler, A. S., Rajulu, D. T., Ronsani, A. E., & Fredrick, B. L. (2008). Peer reviewed: Assessing retail fruit and vegetable availability in urban and rural underserved communities. *Preventing Chronic Disease, 5*(4), 1–9.

Johnston, J. (2008). The citizen-consumer hybrid: Ideological tensions and the case of Whole Foods Market. *Theory and Society, 37*(3), 229–270.

Johnston, J., Szabo, M., & Rodney, A. (2011). Good food, good people: Understanding the cultural repertoire of ethical eating. *Journal of Consumer Culture, 11*(3), 293–318.

Kaufman, P. R. (1999). Rural poor have less access to supermarkets, large grocery stores. *Rural Development Perspectives, 13*(3), 19–26.

Keleman, A. (2009). Institutional support and in situ conservation in Mexico: Biases against small-scale maize farmers in post-NAFTA agricultural policy. *Agriculture and Human Values, 27*(1), 13–28.

Khoury, C. K., Bjorkman, A. D., Dempewolf, H., Ramirez-Villegas, J., Guarino, L., Jarvis, A.,... Struik, P. C. (2014). Increasing homogeneity in global food supplies and the implications for food security. *Proceedings of the National Academy of Sciences, 111*(11), 4001–4006.

Larimore, S. H. (2014). *Boundaries to access in Charlotte, NC farmer's markets accepting supplemental nutrition assistance program* (Unpublished master's thesis). University of North Carolina Charlotte, NC.

Levay, A. V., Mumtaz, Z., Rashid, S. F., & Willows, N. (2013). Influence of gender roles and rising food prices on poor, pregnant women's eating and food provisioning practices in Dhaka, Bangladesh. *Reproductive Health, 10*, 1–11.

Liese, A. D., Weis, K. E., Pluto, D., Smith, E., & Lawson, A. (2007). Food store types, availability, and cost of foods in a rural environment. *Journal of the American Dietetic Association, 107*(11), 1916–1923.

Lobao, L. M., Schulman, M. D., & Swanson, L. E. (1993). Still going: Recent debates on the Goldschmidt Hypothesis. *Rural Sociology, 58*(2), 277–288.

Lobao, L., & Meyer, K. (2001). The great agricultural transition: Crisis, change, and social consequences of twentieth century U.S. farming. *Annual Review of Sociology, 27*, 103–124.

Lyson, H. C. (2014). Social structural location and vocabularies of participation: Fostering a collective identity in urban agriculture activism. *Rural Sociology, 79*(3), 310–335.

Macias, T. (2008). Working toward a just, equitable, and local food system: The social impact of community-based agriculture. *Social Science Quarterly, 89*(5), 1086–1101.

Marktanner, M., & Noiset, L. P. (2013). Food price crisis, poverty, and inequality. *The Developing Economies, 51*(3), 303–320.

Martinez-Gomez, F., Aboites-Manrique, G., & Constance, D. H. (2013). Neoliberal restructuring, neoregulation, and the Mexican poultry industry. *Agriculture and Human Values, 30*(4), 495–510.

Mazzeo, J. (2009). *Lavichè*: Haiti's vulnerability to the global food crisis. *NAPA Bulletin, 32*(1), 115–129.

McIntyre, L., Glanville, N. T., Raine, K. D., Dayle, J. B., Anderson, B., & Battaglia, N. (2003). Do low-income lone mothers compromise their nutrition to feed their children? *Canadian Medical Association Journal, 168*(6), 686–691.

Morris, P. M., Neuhauser, L., & Campbell, C. (1992). Food security in rural America: A study of the availability and costs of food. *Journal of Nutrition Education, 24*(1), 52S–58S.

Morton, L. W., & Blanchard, T. C. (2007). Starved for access: Life in rural America's food deserts. *Rural Realities, 1*(4), 1–10.

Munoz-Plaza, C. E., Filomena, S., & Morland, K. B. (2007). Disparities in food access: Inner-city residents describe their local food environment. *Journal of Hunger & Environmental Nutrition, 2*(2/3), 51–64.

Peters, D. J. (2002). *Revisiting the Goldschmidt hypothesis: The effect of economic structure on socioeconomic conditions in the rural Midwest* (Technical Paper P-0702–1). Jefferson City: Missouri Economic Research & Information Center (MERIC).

Pollan, M. (2003, October 12). The (agri)cultural contradictions of obesity. *New York Times*. Retrieved from http://www.nytimes.com/2003/10/12/magazine /12WWLN.html

Popkin, B. M., Duffey, K., & Gordon-Larsen, P. (2005). Environmental influences on food choice, physical activity and energy balance. *Physiology & Behavior, 86*(5), 603–613.

Powell, L. M., Slater, S., Mirtcheva, D., Bao, Y., & Chaloupka, F. J. (2007). Food store availability and neighborhood characteristics in the United States. *Preventive Medicine, 44*(3), 189–195.

Raja, S., Ma, C., & Yadav, P. (2008). Beyond food deserts: Measuring and mapping racial disparities in neighborhood food environments. *Journal of Planning Education and Research, 27*(4), 469–482.

Rogers, J., & Fraszczak, M. (2014). "Like the stem connecting the cherry to the tree": The uncomfortable place of intermediaries in a local organic food chain. *Sociologia Ruralis, 54*(3), 321–340.

Ruel, M. T., Garrett, J. L., Morris, S. S., Maxwell, D. G., Oshaug, A., Engle, P. L.,...Haddad, L. J. (1998). *Urban challenges to food and nutrition security: A review of food security, health, and caregiving in the cities.* Washington, DC: International Food Policy Research Institute (IFPRI). Retrieved from http://www.ifpri.org/sites/default/files/publications/dp51.pdf

Schnell, S. M. (2013). Food miles, local eating, and community supported agriculture: Putting local food in its place. *Agriculture and Human Values, 30*(4), 615–628.

Schofer, E., & Meyer, J. W. (2005). The worldwide expansion of higher education in the twentieth century. *American Sociological Review, 70*(6), 898–920.

Short, A., Guthman, J., & Raskin, S. (2007). Food deserts, oases, or mirages? Small markets and community food security in the San Francisco Bay area. *Journal of Planning Education and Research, 26*(3), 352–364.

Smith, L. C., El Obeid, A. E., & Jensen, H. H. (2000). The geography and causes of food insecurity in developing countries. *Agricultural Economics, 22*(2), 199–215.

Stuff, J. E., Horton, J. A., Bogle, M. L., Connell, C., Ryan, D., Zaghloul, S., . . . Szeto, K. (2004). High prevalence of food insecurity and hunger in households in the rural Lower Mississippi Delta. *Journal of Rural Health, 20*(2), 173–180.

Tickamyer, A. R., & Duncan, C. M. (1990). Poverty and opportunity structure in rural America. *Annual Review of Sociology, 16*, 67–86.

Tolbert, C. M., & Lyson, T. A. (1992). Earnings inequality in the nonmetropolitan United States: 1967–1990. *Rural Sociology, 57*(4), 494–511.

Trubek, A. B., & Bowen, S. (2008). Creating the taste of place in the United States: Can we learn from the French? *GeoJournal, 73*(1), 23–30.

United Nations (UN). (2011). *The global social crisis: Report on the world social situation 2011.* New York, NY: United Nations.

Warde, A. (2000). Eating globally: Cultural flows and the spread of ethnic restaurants. In D. Kalb, M. van der Land, R. Staring, B. van Steenbergen, and N. Wilterdink (Eds.), *The ends of globalization: Bringing society back in* (pp. 299–316). Washington, DC: Rowman & Littlefield.

Weber, K., Heinze, K. L., & DeSoucey, M. (2008). Forage for thought: Mobilizing codes in the movement for grass-fed meat and dairy products. *Administrative Science Quarterly, 53*(3), 529–567.

Wilkins, J. L. (2005). Eating right here: Moving from consumer to food citizen. *Agriculture and Human Values, 22*(3), 269–273.

Yeh, M., Ickes, S. B., Lowenstein, L. M., Shuval, K., Ammerman, A. S., Farris, R., & Katz, D. L. (2008). Understanding barriers and facilitators of fruit and vegetable consumption among a diverse multi-ethnic population in the USA. *Health Promotion International, 23*(1), 42–51.

Zenk, S. N., Schulz, A. J., Israel, B. A., James, S. A., Bao, S., & Wilson, M. L. (2005). Neighborhood racial composition, neighborhood poverty, and the spatial accessibility of supermarkets in metropolitan Detroit. *American Journal of Public Health, 95*(4), 660–667.

Zenk, S. N., Odoms-Young, A. M., Dallas, C., Hardy, E., Watkins, A., Hoskins-Wroten, J., & Holland, L. (2011). "You have to hunt for the fruits, the vegetables": Environmental barriers and adaptive strategies to acquire food in a low-income African American neighborhood. *Health Education & Behavior, 38*(3), 282–292.

Zezza, A., & Tasciotti, L. (2010). Urban agriculture, poverty, and food security. Empirical evidence from a sample of developing countries. *Food Policy, 35*(4), 265–273.

Chapter 9

The Death of Distance: Food Deserts Across the Global Divide

Meredith Gartin

Introduction

The twenty-first century has been referred to as the *urban century*. More than half of humanity lives in cities (Boone & Modarres, 2006). The majority of cities are unable to produce enough food to sustain their residents (Pothukuchi & Kaufman, 1999). However, some cities are exploring the idea that urban agriculture can create a more resilient food system (Grewal & Grewal, 2012). Resilience of a food system implies that regardless of any external forces (e.g., climate change, hazard event, or economic crisis) to food production or supply systems, residents remain food secure until the system is restored (Grewal & Grewal, 2012). Yet, some argue that the global food system is too interconnected with our local food systems and that no system can exist without some level of vulnerability to external forces (Evers, 1994; Plattner, 1985; Pottier, 1999). The purpose of this chapter is to focus on three case studies that explore the idea of these interlocking food deserts to illustrate how global and local processes converge on health issues (such as obesity) in urban settings.

Obesity Epidemic

Overweight and *obesity* are defined as an abnormal or excessive fat accumulation that impairs health. The World Health Organization reports that obesity has doubled since 1980, and around 3.4 million adults die each year as a result of being overweight or obese (WHO, 2009). Body mass index (BMI) is a measure of weight-for-height that is used to classify overweight and obesity in adults. It is defined as a person's weight in kilograms, divided by the square of their height in meters (kg/m^2); a measure of twenty-five to less than thirty is classified as overweight, and greater than thirty is obese. Approximately one in every three adults is classified as overweight and one in every nine as obese (Stevens, et al., 2012; WHO, 2009).

Understanding the obesity epidemic requires an acceptance of a well-documented fact: humans live in a world that is interconnected (Robertson, 1992). Adam Drewnowski and Barry M. Popkin (1997) identified evidence of a nutrition transition in the 1990s that explained the increasing rates of obesity in developing countries. Before their discovery, obesity was considered an epidemic of the "West," or was associated with affluence. The nutrition transition explained a dietary shift from a traditional diet to a Western diet characterized by obesogenic (calorically-dense) food. So, obesity exists in a complex, "global" system as an epidemic (Popkin, 2006; Popkin & Gordon-Larsen, 2004).

Various household strategies and individual choices were assumed to account for the spread of obesity (Brewis & Gartin, 2006). Interventions at the individual level often recommend that people limit energy intake from total fats and sugars, increase consumption of fruit and vegetables—as well as legumes, whole grains, and nuts—and engage in regular physical activity (60 minutes a day for children and 150 minutes per week for adults). Recent obesity perspectives, however, suggest that the burden may not fall on the individual solely but also on the design and structure of community interactions and structures in the local food systems (Christakis & Fowler, 2007).

As obesity continues to climb in populations around the world, researchers continue to address whether obesity is an individual problem, a community problem, or both. Are people getting sick and dying from obesity because they make poor dietary choices, or because their choices are limited by their community's food system or city's services? How do food environments become limited in food supplies, and how can we prevent such disparities from emerging as cities continue to grow? Examining food deserts allows researchers to explore these questions and to understand the experience of living in a food desert, while seeking strategies to improve access to healthy food. If people live in a place that is limited in healthy and affordable food, then the solutions to this problem would appear to be place-based.

Food Deserts as Place-Based Crises

Food deserts are characterized by socioeconomic status (SES) and significant food environment indicators such as walkability, availability, affordability, and quality of local food stores (Hemphill, Raine, Spence, & Smoyer-Tomic, 2008; Inglis, Ball, & Crawford, 2008; Latham & Moffat, 2007; Macintyre, Macdonald, & Ellaway, 2008). Typically, supermarkets are considered the best source of nutrition when compared to other types of food stores, because they are larger and offer more fresh produce and healthy options at affordable prices (Cummins et al., 2009; Freedman & Bell, 2009; Glanz, Sallis, Saelen, & Frank, 2007). In a global context, however, the informality of street and open-air markets in less developed countries may provide a wider array of stores and greater competition to keep prices low. Either way, social scientists are still learning how urban residents navigate their food environment in search of healthy, affordable, and fresh food, while policy makers are tasked with the job to narrow this gap to create more equitable and service-providing food environments.

The following sections describe three food desert case studies. The first is cited as one of the original food deserts: Springburn in the United Kingdom. The second case study is less well known: Cleveland, Ohio, in the United States, which shares a similar story with other (more well-known) US food deserts, where food deserts are located in historically segregated, blue-collar neighborhoods (Gallagher, 2006). Cleveland is unique, however, because it is considered to have the one of the most progressive urban agricultural policies and movements in the country (Smith, 2012). Finally, the third case study focuses on dietary behaviors in a Paraguayan food desert. Before 2012, researchers implied that food deserts could exist in the global South; yet no prior empirical evidence provided support for food deserts until the publication of a food desert in South Africa (Battersby, 2012) and in San Lorenzo, Paraguay (Gartin, 2012a; 2012b).

Springburn, Glasgow (Scotland)

Springburn, Glasgow, in Scotland, is considered one of the most deprived urban areas in Scotland (Cummins, Findlay, Petticrew, & Sparks, 2005). For Springburn residents, the average income is about one-third below the national average, and they have lower levels of fruit and vegetable consumption and higher levels of poor health, including obesity (NHS Health Scotland, 2004). Cummins and Macintyre (2002) explain that "the major food retailers are held partly responsible for the emergence of food deserts [in Springburn] for not establishing shops in poor communities and so denying residents the benefits of choice and a good price" (p. 436).

Leading retailers recognized that there were development opportunities for them in places like Springburn (Cummins, Findlay, Petticrew, & Sparks, 2005), because not only would larger supermarket stores provide better access to food but they could also act as catalysts for economic development (Cummins & Macintyre, 2002). Of all the large food retailers in the United Kingdom, British Tesco established the Tesco Regeneration Partnership with food desert researchers and practitioners to reclaim food security in UK food deserts (Cummins, Findlay, Petticrew, and Sparks, 2005). The development of a Tesco store in Springburn started in November 1999.

Before opening the Tesco store in Springburn, surveys were conducted with 3,975 households. Multiple follow-up surveys were also conducted to evaluate changes in shopping and dietary behaviors that resulted from increased access to the Tesco store. In 2005, Cummins, Petticrew, Higgins, Findlay, and Sparks published the first set of follow-up surveys and revealed that residents who switched their shopping to the new supermarket improved their self-reported health, meaning that residents felt better about shopping for food.

Then, in 2007, Petticrew, Cummins, Sparks, and Findlay published a second follow-up study. This time, all the residents who had switched their shopping to the supermarket significantly improved their fruit and vegetable consumption. Shoppers who switched also reported an increase in the mean disposable income; before the development of a Tesco store, the average weekly disposable income was less than £100, whereas now, it was more than £120 (Cummins, Findlay, Petticrew, & Sparks, 2005). From 2000 to 2003, there was a 25 percent increase in retail-sector jobs, and not all of those were solely associated with the Tesco development itself (Cummins, Findlay, Petticrew, & Sparks, 2005; Petticrew, Cummins, Sparks, & Findlay, 2007). In addition, the vitality of the retail market saw a reduction in vacant lots, with the construction of new houses and two new secondary school buildings (Cummins, Findlay, Petticrew, & Sparks, 2005).

Cleveland, Ohio (United States)

Cleveland, Ohio, like other post-industrial cities in the United States, has identified a number of food deserts in its most deprived and economically disadvantaged neighborhoods. By one estimate, Northeast Ohioans annually spend $7 billion on food; Clevelanders accounting for about $1 billion of that number, and less than 5 percent of the food is produced locally (Sterpka, 2009). Most food is shipped to

Cleveland from far away—an average trip of 1,300 miles from farm to plate (Sterpka, 2009). Consequently, billions of dollars flow out of the local economy and into industrial conglomerates. The city of Cleveland began to consider that if Northeast Ohio could grow its food locally, then hundreds of millions of dollars could be captured locally (Flachs, 2010; Sterpka, 2009). Today, Cleveland's urban agriculture policy is considered one of the most progressive in the country (Smith, 2012; Sterpka, 2009).

Instead of forming partnerships with large food retailers, the city of Cleveland adopted an amendment in 2007 to establish an Urban Garden District (Cleveland Ohio Code of Ordinances § 336.01). The amendment permitted the use of a wide variety of farming technologies and strategies (e.g., greenhouses, hoophouses, cold-frames, open space, compost bins, chicken coops, beehives, etc.) and other aesthetic features to enhance neighborhoods (e.g., fences, signs, benches, bike racks, raised beds, garden art) (Cleveland Ohio Code of Ordinances § 336.02–336.04). In addition, the City Department of Economic Development sponsors "Gardening for Greenbacks"—"a program designed to ensure that city residents have access to fresh food. Grants up to $3,000 are available to eligible urban farmers for the purchase of tools, irrigation equipment, and greenhouses" (Smith, 2012). As a result of this food renaissance, Cleveland now has over 225 community gardens and, as of 2012, had more farmers markets per capita than any other US metro area (Smith, 2012).

For example, Neighborhood Progress Inc. (NPI), a funding intermediary in Cleveland, has worked closely with community groups since 2008 to create gardens to feed people while building community and strengthening markets to counteract the negative impacts of population loss and abandonment. The mission of NPI is "to foster communities of choice and opportunity throughout Cleveland" (NPI, 2015); and as part of their vision, NPI champions the ReImagining Cleveland initiative—a nationally recognized model for reusing vacant land to create productive landscapes and increase access to healthy, inexpensive food. Lilah Zautner, Director of Sustainability and Urban Greening, playfully stated: "We create gardens and farms as places for people to access healthy, low cost or free food. Access to nutrition alone is a worthy investment; however, when the co-benefits of creating gardens and farms are considered, the investment pencils out. These places provide environmental services. They capture storm water and carbon and provide the needed habitat for the birds and the bees to conduct their very important business... It's only when you witness the wraparound impacts that these places make on the neighborhoods around them, do you begin to see the depth of their value" (see Gartin, 2013).

However, the adoption of urban agricultural products into local households has been slow; lack of money and/or social stigma prevents local residents from accessing the urban farms. Thus, community organizers are tasked with the job to integrate community food sources into the mainstream food market. Although most of Cleveland's urban farms are significantly lower-cost than supermarkets (Flachs, 2010), residents are concerned about the food's origin and safety.

In 2013, I conducted fieldwork in Cleveland. I spoke with community organizers, urban farmers, policy makers, and local funding agencies. They explained that only 5 percent of the at-risk (low-income, food insecure) people are being reached by these gardens. Most Clevelanders will seek out free food programs, of which there are many around town, because unemployment and poverty is a chronic issue. As a result, many farmers are donating their food to food banks, contributing to community meals and potlucks, and recruiting volunteers to work for food (Flachs, 2010). The goal is to transform the ways that people understand and utilize fresh products grown in Cleveland and in surrounding areas.

San Lorenzo, Departamento Central (Paraguay)

San Lorenzo, Paraguay, is the third-largest city in Paraguay (with approximately 250,000 people) and serves as a bedroom community to the nation's capital, Asunción. Before 1990, San Lorenzo served as the primary urban agricultural producer for the capital city. In 1990, however, Paraguay's dictator was overthrown, and the new government signed a treaty agreement with Argentina, Brazil, and Uruguay to create MERCOSUR (*Mercado Común del Sur*—an open, free-trade agreement among countries in the southern cone of South America). The signing of this treaty paved the way for the development of supermarkets and for the replacement of local food produce and commodities with international competitors.

In 2009 and 2010, I conducted ethnographic fieldwork in San Lorenzo (Gartin 2012b). I measured the food environment in terms of price, variety, and quality between various food retailers, including supermarkets, smaller grocers, independent retailers, street vendors, and open-air market stall vendors. The variation between these stores demonstrated that supermarkets—though they offer more food varieties—fail to offer acceptable and fresh produce, whereas the open-air market offered competitively priced fresh and acceptable produce (Gartin, 2012a). I saw, for example, whole bins of rotten produce in supermarkets, priced at the market value, whereas vendors threw out rotten foods in the open-air market. Results also indicated that independent shop owners and smaller grocers scored higher than did supermarkets (Gartin, 2012a). This is likely because the source of their food and produce was the open-air market.

In San Lorenzo, the smaller store owners act as mediators between the area classified as a food desert and the downtown market district. Like the other case locations, supermarkets have developed downtown, leaving the neighborhoods on the periphery without equal access to fresh food. To close the distance between home and market, smaller store owners also have established relationships (consumer partnerships) with their neighbors, who struggle to obtain transportation downtown or who lack employment and job security. This type of partnership allows for residents to purchase food on store credit. Purchasing food on credit is considered a type of coping strategy and is common in other lower-income countries to help keep food supplies available in the home and community (Davies, 1993; Dufour, Staten, Reina, & Spurr, 1997; Maxwell, 1996).

Dietary and nutritional assessments with residents in the food desert found that people who lived longer (more exposed) in San Lorenzo are at a reduced risk for obesity because they are more likely to establish a relationship with local store owners or market vendors. These relationships require a lot of time to establish and to build trust, but residents who shopped at the smaller stores and open-air market were found to eat more fresh fruit, while those who shopped at the supermarket were found to eat more prepared meals and obesogenic snacks (Gartin, 2012b).

In general, people felt secure that food supplies were available; but they feared economic insecurity, unemployment, and the inability to purchase food in the local marketplace. Residents stated that at present they would be able to find food but were uncertain about being able to do so in the future. In the study field site and food desert, a small mural was painted on a pillar in one of the public plazas. Translated, it read: "When we've cut down the last tree/ when we've contaminated the last river/ when we've killed the last fish/ you'll find that you can't eat money" (see Gartin, 2014). The quote captures the concern articulated to me by most residents, regardless of their current job security.

Case Study Highlights

Often, supermarkets are considered the best source of nutritious and affordable foods (Cummins et al., 2009; Freedman & Bell, 2009; Glanz et al., 2007). Supermarkets facilitate the sale of local food and narrow the physical distance between agricultural producers and urban shoppers around the world; however, these three case studies challenge the assumptions attached to supermarkets in urban food environments. Table 9.1 highlights the solutions and results derived from the three case studies in this chapter.

In each case study, supermarkets (or the lack of supermarkets) structure the kinds of access urban residents have in their food environments. In the Springburn (UK) case, for example, the development of a supermarket positively influenced local diet and food security. Additionally, the presence helped revitalize the community in ways that improved economic stability overall. In the UK, supermarkets fill a void left by policies that promoted the development of supermarkets located on the peripheries of cities, leaving downtown and city-central residents without access.

Likewise, Cleveland also strives to fill a void left by inadequate development policies while promoting economic growth. Their approach, however, reflects how city municipalities and local communities are adjusting their development philosophies to encourage both economic and environmental sustainability. Cities around the world are showing that they can unravel, exhibit limited capacities, and become more vulnerable to climate change, economic downturns, and other kinds of global pressure. Cleveland is not only interested in increasing access to healthy and fresh food but also in the creation of

Table 9.1 Results by food desert case studies

City	Solution	Result
Springburn	Development of a research partnership with large food retailers	Significant increase in fruits and vegetables after two years; increased overall food security
Cleveland	Creation of an urban garden district and community organization around urban agriculture	Significant social, environmental, and economic changes; slow adoption of local fresh agriculture and dietary change
San Lorenzo	Residents opening smaller shops to resell food products from the open-air markets in the food desert	Improved community food security; reduced risk of undernutrition and obesity

vibrant social and economic spaces in which capital can cycle within the local community. (A chain supermarket would likely detract from this planning objective.) As a result, the city of Cleveland has enacted new policies to encourage sustainable urban farming at multiple levels (households, neighborhoods, city, and region).

Yet only a small percentage of the at-risk population in Cleveland utilizes fresh and local food markets. The change in cultural and social habits to shop at a neighbor's farm stand versus a large chain retail store has been slow to adopt. While I was in Cleveland, a community organizer also explained to me that residents still choose to spend what little cash they have on bus transportation to access "free" city food programs. Even though the market food is low-cost, the organizers and farmers have not been able to reach as many people as they identify as being in-need or at-risk. Therefore, community organizers and farmers are offering some free food and community meals that are supported by local agricultural producers to emphasize their presence as a viable and economical food source.

Supermarkets in San Lorenzo, on the other hand, negatively impact the city. In Latin America, the supermarket sector is mostly foreign-owned and is undergoing rapid concentration in cities. British Tesco, French Carrefour and Casino, Dutch Ahold and Makro, Belgian Food Lion, and United States Walmart are all top global multinational retailers and account for 65 percent of all global food sales (Reardon, Timmer, & Berdegué, 2005). Retail supermarkets feature low prices that are competitive with family-run businesses and street markets (Reardon et al., 2005). The only caveat is that the replacement of traditional markets with supermarkets has been slower in the fresh-produce sector of food sales (opposite from Cleveland).

Despite the supermarkets' large share of produce sales, local agriculture production still finds its way into the marketplace. Processed foods, canned produce, meats/dairy, and other types of food products have replaced a good percentage of food sales, but local and fresh produce remains a thriving retail sector in the developing world. In Brazil, for example, fruits and vegetables from supermarkets constitute about 50 percent of the fresh-produce sales, though produce was primarily purchased outside supermarkets during the 1980s; while supermarkets in Argentina garner around 60 percent of the total retail share, they sell 30 percent of the country's fresh produce (Reardon et al., 2005).

Though other food products found in supermarkets have seen faster growth in sales, fresh-produce sales have increased significantly as well. Fresh-fruit consumption in Paraguay, for example, was found most significant among shoppers at the neighborhood store or in the open-air market, while shoppers at the supermarket consumed more prepared meals. I observed, while conducting ethnographic research, that prepared meals are laid out like a buffet in supermarkets; residents fill a plate or a box with any assortment of food, and the plate is weighed and priced by the kilo. Quite often, portions were large.

Also, I observed in the field, that produce trucks from Brazil and Argentina would stop at the supermarkets and the open-air market. Street and market vendors explained that some produce items are exclusively from Brazil or Argentina and that Paraguayans no longer grow their own produce. (Tomatoes in particular are rarely from Paraguay but are used every day for cooking.) Market vendors said that local producers can't compete with the international prices. Competitive prices are helping supermarkets make inroads into increasing the sales of fruits and vegetables internationally (Reardon et al., 2005).

Supermarkets have benefited from government regulations that are intended to improve urban congestion and sanitation problems as the regulations have posed more challenges to open-air markets (Reardon et al., 2005). As these street markets struggle to compete, it becomes clear that the food environment is rapidly changing and that food cultures are transitioning toward more obesogenic foods.

Implications

Are we (the West) exporting our food environments along with our food products? As global multinational food retailers carve their stores into the city infrastructure, how does their presence impact local diets and food cultures? Under what conditions do supermarkets encourage urban renewal versus the conditions that encourage exclusion? How can policy makers and concerned citizens of less developed countries establish new paths toward more resilient food systems, considering the knowledge we currently have from developed countries? Gaining answers to these questions requires science and society to consider the connections that tie local communities to global economies.

The design of our global food system has effectively drawn countries with very little capital into a network of food trade. Paraguay, for example, remained isolated from global competition until the signing of the trade agreement MERCOSUR. Now, a large majority of everyday produce is shipped in from Brazil and Argentina. The death of distance that has occurred in the last decades with increased food trade has resulted in efficient passageways that carry food from far away. The current food system is less constrained by space and physical distance around the world; however, this does not mean that our connections around the world are stable.

Environmental hazards and climatic events (e.g., floods, fires, and droughts) can reduce food supplies in one country and affect the supplies in another. Even economic crashes in one country can affect our global food system. In 2008 and 2011, the Global Food Crisis drew attention to the political and economic forces structuring market food prices. The executive director of the World Hunger Program said: "We're seeing more people hungry and at greater numbers than before. There is food on the shelves, but people are priced out of the market" (Holt-Giménez & Peabody, 2008). Thus, in the case of the global food crisis, the "death of distance" resulted in the exacerbation of economic disparities between high- and low-income people and high- and low-income countries.

Conclusions and Observations

Food deserts form out of community and international development policies. The intent of these policies is to design cities in ways that provide human services to fill basic needs. Food as a commodity holds power because we all need it to survive. However, for one group (such as a corporation) to benefit at the expense of another group is unjust. Examining and understanding food deserts as a type of food environment allows researchers and practitioners to examine how food as a commodity flows from its origin—where the food was produced—to the urban marketplace and into individual households. This allows researchers to examine topics such as food rights and city policies, as well as nutritional health and global epidemics.

Food shapes our health as well as our identities and environments. Given the rapid growth of supermarkets and their increased concentration in less developed countries, it is important that our research and policy work toward finding solutions to improve food security at the point where global and local processes converge, in urban food environments, in particular.

Acknowledgments

This material is based upon work supported by the National Science Foundation under Grant No. RCN 1140070. Any opinions, findings, and conclusions or recommendations expressed in this material are those of the author and do not necessarily reflect the views of the National Science Foundation.

References

Battersby, J. (2012). Beyond the food desert: Finding ways to speak about urban food security in South Africa. *Geografiska Annaler: Series B, Human Geography, 94*(2), 141–159.

Boone, C. G., & Modarres, A. (2006). *City and environment.* Philadephia, PA: Temple University Press.

Brewis, A., & Gartin, M. (2006). Biocultural construction of obesogenic ecologies of childhood: Parent-feeding versus child-eating strategies. *American Journal of Human Biology, 18*(2), 203–213.

Christakis, N. A., & Fowler, J. H. (2007). The spread of obesity in a large social network over 32 years. *New England Journal of Medicine, 357*(4), 370–379.

Cummins, S., Findlay, A., Petticrew, M., & Sparks, L. (2005). Healthy cities?: The impact of food retail led regeneration on food access, choice and retail structure. *Built Environment, 31*(4), 288–301.

Cummins, S., & Macintyre, S. (2002). "Food Deserts"—evidence and assumption in health policy making. *British Medical Journal, 325*(7361), 436–438.

Cummins, S., Petticrew, M., Higgins, C., Findlay, A., & Sparks, L. (2005). Large scale food retailing as an intervention for diet and health: Quasi-experimental evaluation of a natural experiment. *Journal of Epidemiology and Community Health, 59*(12), 1035–1040.

Cummins, S., Smith, D. M., Taylor, M., Dawson, J., Marshall, D., Sparks, L., & Anderson, A. S. (2009). Variations in fresh fruit and vegetable quality by store type, urban-rural setting and neighbourhood deprivation in Scotland. *Public Health Nutrition, 12*(11), 2044–2050.

Davies, S. (1993). Are coping strategies a cop out? *IDS Bulletin, 24*(4), 60–72.

Drewnowski, A., & Popkin, B. M. (1997). The nutrition transition: New trends in the global diet. *Nutrition Reviews, 55*(2), 31–43.

Dufour, D. L., Staten, L. K., Reina, J. C., & Spurr, G. B. (1997). Living on the edge: Dietary strategies of economically impoverished women in Cali, Colombia. *American Journal of Physical Anthropology, 102*(1), 5–15.

Evers, H-D. (1994). The trader's dilemma: A theory of the social transformation of markets and society. In H-D. Evers & H. Schrader (Eds.), *The moral economy of trade: Ethnicity and developing markets* (pp. 68–75). London: Routledge.

Flachs, A. (2010). Food for thought: The social impact of community gardens in the greater Cleveland area. *Electronic Green Journal, 1*(30). Retrieved from https://escholarship.org/uc/item/6bh7j4z4

Freedman, D. A., & Bell, B. A. (2009). Access to healthful foods among an urban food insecure population: Perceptions versus reality. *Journal of Urban Health, 86*(6), 825–838.

Gallagher, M. (2006). *Examining the impact of food deserts on public health in Chicago.* Retrieved from http://www.lasallebank.com/about/pdfs/report.pdf

Gartin, M. (2012a). Food deserts and nutritional risk in Paraguay. *American Journal of Human Biology, Special Issue on Global Obesity, 24*(3), 296–301.

Gartin, M. (2012b). *Residence in a deprived food environment: Food access, affordability, and quality in a Paraguayan food desert.* PhD Diss., Arizona State University.

Gartin, M. (2013, March). Gardens feed and fuel grassroot organization: Neighborhood Progress Inc. *Anthropology News, 54.*

Gartin, M. (2014). Op-ed: Money can't be eaten. *Tvergastein Interdisciplinary Journal of the Environment, Special Issue on Food Fights and Food Rights,* in press at http://tvergasteinjournal.wordpress.com/archive/

Glanz, K., Sallis, J. F., Saelens, B. E., & Frank, L. D. (2007). Nutrition environment measures survey in stores (NEMS-S): Development and evaluation. *American Journal of Preventive Medicine, 32*(4), 282–289.

Grewal, S. S., & Grewal, P. S. (2012). Can cities become self-reliant in food? *Cities, 29*(1), 1–11.

Hemphill, E., Raine, K., Spence, J. C., & Smoyer-Tomic, K. E. (2008). Exploring obesogenic food environments in Edmonton, Canada: The association between socioeconomic factors and fast-food outlet access. *American Journal of Health Promotion, 22*(6), 426–432.

Holt-Giménez, E., & Peabody, L. (2008). From food rebellions to food sovereignty: Urgent call to fix a broken food system. *Institute for Food Development and Policy, 14*(1), 1–6.

Inglis, V., Ball, K., & Crawford, D. (2008). Socioeconomic variations in women's diets: What is the role of perceptions of the local food environment? *Journal of Epidemiology and Community Health, 62*(3), 191–197.

Latham, J., & Moffat, T. (2007). Determinants of variation in food cost and availability in two socioeconomically contrasting neighbourhoods of Hamilton, Ontario, Canada. *Health & Place, 13*(1), 273–287.

Macintyre, S., Macdonald, L., & Ellaway, A. (2008). Do poorer people have poorer access to local resources and facilities? The distribution of local resources by area deprivation in Glasgow, Scotland. *Social Science and Medicine, 67*(6), 900–914.

Maxwell, D. G. (1996). Measuring food insecurity: The frequency of severity of "coping strategies." *Food Policy, 21,* 291–303.

Neighborhood Progress Inc. (NPI). (2015). Cleveland neighborhood progress: Mission / vision. Retrieved from http://www.npi-cle.org/about/mission-vision/

NHS Health Scotland. (2004). *Constituency health and well-being profiles 2004—Springburn, Shettleston.* Retrieved from www.healthscotland.com/profiles

Petticrew, M., Cummins, S., Sparks, L., & Findlay, A. (2007). Validating health impact assessment: Prediction is difficult (especially about the future). *Environmental Impact Assessment Review, 27*(1), 101–107.

Plattner, S. (1985). Equilibrating market relations. In S. Plattner (Ed.), *Markets and marketing* (pp. 133–152). Lanham, MD: University Press of America Monographs in Economic Anthropology.

Popkin, B. M. (2006). Global nutrition dynamics: The world is shifting rapidly toward a diet linked with noncommunicable diseases. *American Journal of Clinical Nutrition, 84*(2), 289–298.

Popkin, B. M., & Gordon-Larsen, P. (2004). The nutrition transition: Worldwide obesity dynamics and their determinants. *International Journal of Obesity, 28*(Suppl 3), S2–S9.

Pothukuchi, K., & Kaufman, J. L. (1999). Placing the food system on the urban agenda: The role of municipal institutions in food systems planning. *Agriculture and Human Values, 16*(2), 213–224.

Pottier, J. (1999). *Anthropology of food: The social dynamics of food security.* Cambridge: Polity Press.

Reardon, T., Timmer, C. P., & Berdegué, J. A. (2005). Supermarket expansion in Latin America and Asia: Implications for food marketing systems. In A. Regmi & M. Gehlhar (Eds.), *New directions in global food markets.* Economic Research Service/United States Department of Agriculture, Agriculture Information Bulletin Number 794.

Robertson, R. (1992). *Globalization: Social theory and global culture.* London: Sage Publications.

Smith, J. (2012). Encouraging the growth of urban agricultural in Trenton and Newark through amendments to the zoning codes: A proven approach to addressing the persistence of food deserts. *Vermont Journal of Environmental Law 14*(1), 71–100.

Sterpka, M. (2009). Cleveland's for-profit urban gardens are growing. *Cleveland.com.* Retrieved from http://blog.cleveland.com/metro/2009/07/clevelands_forprofit_urban_gar.html

Stevens, G. A., Singh, G. M., Lu, Y., Danaei, G., Lin, J. K., Finucane, M. M., . . . Ezzati, M. (2012). National, regional, and global trends in adult overweight and obesity prevalences. *Population Health Metrics 10.* Retrieved from http://www.pophealthmetrics.com/content/10/1/22

World Health Organization (WHO). (2009). *Global health risks: Mortality and burden of disease attributable to selected major risks.* Geneva: World Health Organization.

Part IV

The Future of Food

Chapter 10

Place-Based Responses to the Global Food Economy

Andria D. Timmer

Introduction

The global food economy is one based on the creation of shelf-stable foodstuffs, transportation, and mass production. This industrialized system gives consumers the illusion of freedom from the caprices of nature and seasonality and allows fewer people to be engaged in the business of food production. Ostensibly, it allows more people to eat, based on the labor of a few, but the global food economy is not without its drawbacks. In recent years, it has become clear that food produced in such high quantities tends to lack nutritional value and is responsible for the growing health crisis marked by rising rates of obesity, heart disease, and type 2 diabetes. Moreover, food produced in this manner takes an environmental toll. It is estimated that 25 percent of food produced each year is wasted and ends up in landfills. Concentrated agricultural feeding operations (CAFOs) are associated with high energy expenditures, water and air pollution, and improper care of animals (Imhoff, 2010). At the same time, food that is produced en masse is vulnerable to contamination, which is evidenced by a recent upsurge in recalls of meat, vegetable, and nut products contaminated with salmonella and E. coli. The purpose of this chapter, however, is not to critique the global food economy but, rather, to analyze one aspect of the response to "big food." This response is centered on local food and, as a place-based movement, is in direct opposition to the placeless-ness of the industrialized system.

A social movement, broadly termed the food revolution or food movement, is emerging in response to the perceived failure of this global food economy in which food is a commodity, not sustenance for life. The food revolution is a diverse social movement based on the ideology that consumers must be more aware of what they eat and, more importantly, where their food is coming from. There is no one food movement, but rather, a growing number of people are questioning the current food culture and are requesting such things as local food, Slow Food, or a sustainable food supply. They are all grouped together because they share the same "recognition that industrial food production is in need of reform because its social/environmental/public health/animal welfare/gastronomic costs are too high" (Pollan, 2010).

This movement is place-based in that it is grounded in local landscapes. Conversely, the global food system is perceived as being placeless. This is, however, a false distinction. As Weiss (2011) argues, "Industrial eaters inhabit a world that is not null, but, in fact, full of places" (p. 442). Industrial food earns the distinction of being without a place, because it can be eaten out of place and time. The global, industrial agricultural system is one that is based on abstraction in that agriculture is disembedded from its sociological, ecological, and cultural foundations (Thivet, 2014). Vasavi (1994) notes how the global system "largely displaced the local knowledge and autonomy of agriculturalists and substitutes the uniform and market-oriented prescriptions of the bureaucracy" (p. 294). "Place," then, is invoked as the remedy to many social ills including, but not limited to, environmental degradation, public health crises, and exploitation of people and animals (see Weiss, 2011, 2012). Thus, a sustainable food system is by definition locally rooted and dependent on a strong "sense of place." For participants in the food revolution, meats, dairy, fruits, and vegetables should be locally grown, sold, and consumed. By eating local foods, they can have an intimate knowledge of how their food was grown and raised and thus have a connection both to the land and to the farmer. Knowing who grew and raised the food, how and where it was grown, is of paramount importance. For this reason, foods that are considered unhealthy by the mainstream, butter and lard, for example, could be deemed healthy and nourishing *if* produced locally from free-range, grass-fed animals. Where the food comes from is more important in determining the overall healthfulness than its specific nutritional content.

In this chapter, I discuss the importance of place in the growing food movement. I begin with a brief and selective overview of the history and development of the food revolution, of which the move to eat and to buy local (i.e., in place) is a part. There is no one food revolution, as food has often been used as a tool for revolutionary action. As Wilk (2006) points out, "Food has long been a focus for political and social movements in many parts of the world; food is a potent symbol of what ails a society, a way of making abstract ideas such as class or exploitation into a material, visceral reality" (pp. 21–22). Due to the breadth and depth of the food movement writ large, I am, by necessity, leaving out many important players, including Slow Food, fair trade, and food coops. I focus instead on those aspects of the food movement that are clearly rooted in a sense of place. This includes the use of farmers markets, urban and community gardening, and homesteading. I discuss these in sections on the what, why, and how of local food. I then draw upon my own research conducted on the East Coast of Virginia and Maryland to provide an ethnographic example of one group for whom eating locally and in place is of great dietary, sociological, and spiritual significance. The chapter concludes with a discussion of some of the critiques and dilemmas associated with a food movement based on a sense of place.

An Emerging Revolution

The food revolution is growing in popularity, support, and recognition; but it is erroneous to believe that food movements are new. Food has always been a vehicle for social revolutionaries. For example, the names now most commonly associated with breakfast cereal—Sylvester Graham and John Harvey Kellogg—called for a whole food, vegetarian diet in the early nineteenth century. Their goals, however, differed dramatically from those of current food revolutionaries. They did not aim to be local or to fight against a global food system. Rather, they endeavored to cure people of their sexual urges and encourage morality (Whorton, 1994).

The food movement of the 1970s was dominated by activists such as Frances Moore Lappé (1971), who argued for a vegetarian diet to promote individual and environmental health. She encouraged consumers to be aware of the ecological impact of eating, for example, by explaining the resources involved in meat production. She was not, however, overly invested in a turn to the local. In recent years, the rhetoric of the food movement has slowly turned away from *what* food to *where* food. Some key figures in this most recent reincarnation of the movement include chef and activist Alice Waters, urban gardener Will Allen, writer Michael Pollan, and farmer and speaker Joel Salatin. The most engaged activists are the ones who are not in the public eye—the food purchasers who are buying and preparing food for their families. Therefore, although the film *Food, Inc.* and, more recently, *Fed Up,* and the writings of Pollan (2006, 2008,

2009), Salatin (2007, 2011), Allen (2012), and others (Planck, 2006; Robbins, 2011; Schlosser, 2001) lay out the foundation of the movement, it is these private actors who maintain, change, and provide momentum to this movement.

The What and Why of Local

The term *local* is one that belies an easy definition, but place needs boundaries to be actualized. "Local" is as ambiguous a concept as a "sense of place." Both concepts benefit from being grounded in some geographical context. In terms of food, "local" can refer to foodstuffs produced as close as one's backyard or as far as 500 miles away. Some producers use the radial definition and argue that anything grown within a certain radius from the buyer may be considered local. An owner of a Richmond, Virginia–based community supported agriculture (CSA) draws a 150-mile radius around his market, for example, and spends two days a week collecting produce, animal products, and specialty items that fall within this circle. Others use the concept of a region or foodshed (Kloppenburg, Hendrickson, & Stevenson, 1996) where, regardless of mileage, everything grown in a similar region is affected by similar processes and can be considered local. The Chesapeake Watershed, which comprises about two hundred miles of shoreline on the Chesapeake Bay in Virginia and Maryland, is one such region. Supermarkets tend to use state-based criteria. Foods grown in neighboring states, again, regardless of mileage, can be deemed local. Unlike the "organic" label, the "local" label is not regulated by the USDA.

"Local" is open to interpretation. Usually implicit in the idea of local is the sense that food is community-based, sustainable, tastier, and healthier (Janssen, 2010; Weiss, 2011). Those who seek out local food are seeking out a connection with the farmer. Lyson (2012) calls this "civic agriculture," in which local agriculture is embedded in the food production of the local community and this linkage increases "agricultural literacy" (p. 26). This is, of course, directly opposed to industrial agriculture, wherein, as already mentioned, this linkage is disembedded. Born and Purcell (2006), however, warn against getting caught in the "local trap": "the tendency of food activists and researchers to assure something inherent about the local scale" (p. 195). The concept of local brings with it ideas about ecological sustainability, social and economic justice, and even food quality and health; however, local systems produce variable outcomes, and it is not useful to conflate scale with outcome (p. 196).

Regardless of the problems inherent in the local trap, it is evident that the demand for local food is increasing. Farmers and consumers are both involved in challenging the global food complex through alternative food systems that are "immediate, personal and enacted in shared space" (Hinrichs, 2000, p. 295; see also Winter, 2003). Barber (2014a) reports that a majority of Americans now say that sustainability is a priority when purchasing food. Many grocery stores have recently begun to sell produce labeled "locally grown." Farmers markets are encroaching into almost every urban center. Urban gardening in the form of backyard and community gardens is increasing in popularity, and homesteading is becoming a viable option for many. These events indicate that on a relatively large scale, the ideas of locality and a place-based approach to eating are gaining momentum. In a sense, consumers have long wanted to connect with the place-ness of food, and marketers know this. It is for this reason that commercial food producers draw upon images of farm life, orchards, and rurality. A picture of a farmhouse on a gallon of milk obscures how the milk most likely came from an industrialized dairy in which cows are hooked to machines, kept in a constant state of lactation, and packed in extremely close quarters. It is perhaps for this reason that the growth in the local food movement corresponds with increasing information about what happens to the food as it goes from "farm" to table. This includes stories of chickens that are too large and too weak to stand, cows so devastated by mad cow disease that they have to be transported by forklift, and apples sitting in storage a year or more between picking and shipping to the store. As the buying public becomes more informed about the global food economy, they look for solutions to opt out. One solution is a local, place-based movement.

Much of the experience of eating locally is based on the idea of transparency and of knowing where food comes from. Knowing the farmer connects food to a place. Consumers can ask how a product was grown or raised and can be assured that their environmental and moral concerns are met. Small-scale farms that produce the coveted local food often have open-door policies so that potential food purchasers can see how the animals are kept and fed and how fruits and vegetables are organically and sustainably grown. These farms look dramatically different from industrial farms, which are often shielded from view. Joel Salatin is frequently quoted for his remark that the smell of manure is the smell of mismanagement. A well-run farm should not smell offensive. It is interesting to note that Salatin, arguably the most well-known sustainable farmer, is frequently interviewed on his farm, Polyface, as he is engaged in farm-related activities. In

Food, Inc., for example, he is outside with some farmworkers, processing chickens. In the 2011 documentary *Farmageddon*, he is filmed in the pasture, surrounded by his pigs. Such images underscore the importance of land and place and firmly position Salatin as one of the "good guys."

It should be mentioned that purchasers of local food do not always know the farmer. Supermarkets are merely capitalizing on the "buy local" movement, but purchasers of these products are still disconnected from the farmer. However, the local food movement is largely dependent on the development of active social networks in a way that supermarkets are not. Using Putnam's definition of social capital (2000), Glowacki-Dudka, Murray, and Issacs (2013) assert that "the local food movement *requires* the building of social capital through these partnerships to remain viable" (p. 84, emphasis mine). This is true both for the farmer and for the consumer. Direct agricultural production, such as farmers markets and CSA, "promise human connection" (Hinrichs, 2000, p. 295).

This promise of human connection is evident in a high degree of social embeddedness, an economic sociological concept that provides a useful framework to assess spheres of activities and relationships in local markets. Embeddedness refers to the manner in which economic activities are embedded and enmeshed both in economic *and* in non-economic institutions (Polanyi, 1957). In a purely dualist sense, the globalized, *place-less* food system is dis-embedded, whereas the local food economy is rooted in place and embedded (Sonnino, 2007). Direct agriculture markets, such as farmers markets and CSAs, depend on human connections, whereas relations in global food systems are distant and anonymous (Hinrichs, 2000). This, however, is a false dichotomy. Social embeddedness is a factor of *all* economic activity, and, as Hinrichs (2000) points out, direct agricultural markets are fundamentally rooted in commodity relations. Hinrichs joins the concepts of marketness and instrumentalism with embeddedness to provide a more nuanced view of how they function together, ultimately arguing that these three concepts together are critical for understanding local food systems.

The notion of embeddedness serves to highlight the manner in which enhanced social relationships, which are integral to local markets, serve to empower place. That is, food products come to be seen as more socially embedded through "a process of mobilization of values and meanings that construct a place as the 'local'" (Sonnino, 2007, para. 24). It is the perception of an empowered sense of place that adds value to a locally produced product. The place-based social connection a purchaser experiences at a farmers market provides a context within which this consumer comes to perceive and measure, among other things, quality, taste, freshness, and sustainability (see Winter, 2003). Thus, labeling or marketing a product as local makes it a value-added product (de St. Maurice, 2014). In his analysis of pigs in the local food market, Weiss (2011) quotes an informant who confidently asserts that she does not fear competition from pastured pork produced by large-scale distributor Smithfield because "local will always trump any other label" (p. 439).

"Local" has more power than other watchwords like "organic" or "GMO free" because local food is perceived as being less vulnerable to corporate influences. Walmart and Target, for example, have recently jumped onto the organic food bandwagon. Both have pledged to begin offering more natural, organic, and sustainable products. Food activists have reacted to this with skepticism, if not outright fear. Organic food is a hot commodity right now, but the food movement is about transforming food from a capitalistic *commodity* to something that is used to bring health, pleasure, and community. While the decision made by Target and Walmart indicates the growing interest among consumers for sustainability and will lead to a marked shift away from chemically produced and processed foods, it will also lead to a mass production of organic foods, which is in direct opposition to what most active members of the local food movement want. That is, it reifies the placeless-ness, which is at the root of the discontent with the current market.

With large buyers demanding organic food, farmers producing for them will have to work for the prices set by these wholesalers. Efficiency and profitability, rather than integration and sustainability, will come to dominate the system. As Steve McFadden from the Cornucopia Institute asserts, "Farmers are contracted under these values and thereby relegated to the role of corporate vassals, laboring in servitude to fulfill the terms of contract on quantity, quality, timing, and pricing—all factors that have little to do with the rising spirit of the food democracy movement" (McFadden, 2014, n.p.). Activist Vandana Shiva (2000) argues that organic food in the corporate model is not a solution to our problems. Rather, we need to reconsider the manner in which we think about food, farming, and production. For Salatin, all eaters should be able to look out of their windows and see their food. At a college campus talk in February 2014, he urged students to turn the expanses of grass and rooftops into gardens. In this way, the students would be surrounded by, and intimately connected to, the food that they eat. For Salatin, and others like him, growing food locally is a way to not only address environmental concerns by lowering energy costs and decreasing waste, but also to address spiritual concerns. According to Shiva, "Food democracy is about action—changing the way we eat every time we take a bite. It's about people learning, engaging, and acting in our food systems" (McFadden, 2014, n.p.). We cannot act in our food systems if we are removed from them.

A food revolution, therefore, must be grounded in place; but place is also important because food is inherently specific to an environment. Foods are meant to be grown in diverse ecosystems (Salatin, 2011) rather than in homogeneous complexes. They are meant to be grown in particular climates and in specific seasons. Sellers in markets often encourage buyers to purchase "ugly" fruits and vegetables—ones that are misshapen or have blemishes but are still perfectly tasty and healthy. Such products best represent what properly grown food should look like: imperfect and diverse. By seeking out food that is local, eaters are seeking out unadulterated foods.

The How of Local

How does one go about eating locally? The most obvious choice is farmers markets, and, indeed, farmers markets have been growing in popularity and prevalence in the past decade. The USDA (2014) keeps a directory of farmers markets, and, according to their data, the number of markets around the country has increased from 1,755 in 1994 to 8,144 in 2013. Many of the consumers frequenting farmers markets are looking for products that meet certain standards. They ask a lot of questions, and to be competitive in the markets, sellers must meet these standards. The experiences of Karen (a pseudonym), who sells jam at farmers markets in Virginia, illustrate this well (personal communication, 2014). Karen began making jam to sell at fundraisers to raise money for cancer research after her son successfully completed his leukemia treatment. The jam was very popular, and she was encouraged to sell it in other venues. When she first began making the jam, it was with store-bought produce, but when she began to sell at farmers markets, consumers made it clear that they were looking for products made locally from organically grown products. As a response, now all the fruits she uses come from local sources except for the pineapple (used in her Aloha Jam, a popular seasonal favorite). She feels that what sets her apart is that her ingredients are all picked and processed locally. Her jams are extremely popular, and both she and her husband have quit their jobs to dedicate their time to keeping their business alive.

Many "boutique" farmers markets are not at all dependent on local produce or even food products. Some of the smaller markets attempt to capitalize upon the increasing desire for the local, but they cannot draw the sellers either because of their location or because of a questionable consumer base. In these situations, many of the sellers offer artistic items or skin-care products or are resellers. Resellers are not highly regarded in the farmers-market community. They purchase produce from other farms to sell at the market. In one regard, reselling can be necessary. A farmer at a market held in Virginia in May 2013 contended that it was necessary to buy produce from other farms in other states because the customers wanted tomatoes and it was not yet the tomato season in Virginia (personal communication, 2013). Therefore, to make a profit, he needed to find a way to sell tomatoes and other such products. This particular producer was not invited back to the farmers market the following year.

Gardening is a more direct way to eat locally since it cuts out the farmer, who is the middleman. It is for this reason that community and urban gardening has increased in popularity. The benefits of gardening include fresh seasonal produce as well as a reduction in food transportation costs. It is difficult, if not impossible, to quantify the growth in gardening because it can take on so many forms—from container gardening to the forty-acre farm outside Milwaukee, run by Will Allen and his organization, Growing Power. However, it is possible to assess its increase in popularity. Gardening is an integral part of Michelle Obama's Let's Move initiative. According to the USDA (2013), as much as 15 percent of food is grown in urban gardens.

For those who do not want to garden themselves, CSA shares are a good option. The CSA movement began in the United States in the 1980s and has since grown to almost 2,000 operations (Janssen, 2010). Subscribers buy an agricultural share at the beginning of the growing season and then receive a selection of agricultural products throughout it. Producers benefit because they have guaranteed buyers, and payments are front-ended so they receive them when they are most needed. They are able to use subscriber payments to pay for expenses rather than for recouping losses. Subscribers profit because they receive the benefit of local food without doing the growing themselves. They buy directly from the farmer, so the farm-to-table connection is robust. Moreover, they are able to directly connect to season and climate. They learn what grows best in their location at any given time and experience how seasonality directly affects the taste of certain agricultural products. Eggs, milk, and meat all take on different qualities depending upon the food the animal is grazing, so the animal products will also change as the seasons change. This variation is not seen in industrial animal goods fed on a formulated, standard diet.

CSAs are not without their problems. Consumers who are not accustomed to seasonal eating may get frustrated with boxes filled with unfamiliar vegetables. The amount of produce available will vary over the course of the season and subscribers may not get what they want or expect. Upon visiting the farmer who provided him with his local, heritage faro (emmer wheat), Dan Barber realized that much of what was planted on the farm went to waste because the plants that were needed to ensure quality soil were not the specialist crops that consumers wanted. Barber (2014a) writes, "In celebrating the All Stars of the farmers market—asparagus, heirloom tomatoes, emmer wheat—the farm-to-table advocates are often guilty of ignoring a whole class of humbler crops" (n.p.). In his recently released book, Barber (2014b) calls for dining on the "third plate." The "first plate" is the traditional meat-centered diet supplemented by vegetables. The "second plate" is the new ideal of organic, grass-fed meats, and locally grown vegetables. His "third plate" is an integrated system rooted in a celebration of the whole fare. For example, in his restaurant, he offers a rotation risotto made with whatever grains and vegetables are in abundance at the farm at the time. In many ways, CSAs can offer the third plate for which Barber is advocating. However, unfamiliar or repetitive items can often turn off customers; so in order to meet the demands of their customer base, the CSA must provide products that are not from their farm.

Farmers markets and CSAs are about buying local and having a connection to a farmer, and, thus, a place. Homesteading takes this one step further. It is about incorporating the ideals of local into all aspects of life. In modern usage, homesteading does not refer to a government land program, but, rather, to a lifestyle centered on self-sufficiency, which typically involves at-home food production, preparation, and preservation. For example, Abigail (a pseudonym), who is in the process of building her homestead, says that her goal is to not have to leave her home (personal communication, 2014). "Homesteading," writes Gould (2005), a scholar and participant in the movement, "means staying at home, but in the richest possible sense" (p. 2). For Gould, homesteading is a religious movement. It is a way for individuals to make meaning by "choosing to center one's life around home, a home consciously built with attention to a particular place in the natural world" (p. 2). She believes that homesteaders are brought to the movement for spiritual reasons but recognizes that there is great diversity among homesteaders: "For some, homesteading becomes primarily a private, symbolic practice of descent from the dominant culture. For others, it is a means of radically reforming that culture. For still others, it is a delicate balance of both. In all cases, however, homesteading involves not just practical work but also symbolic, cultural work" (2005, p. 5).

Hayes (2010) ascribes to the second position. She advocates for what she calls "radical homemaking": homesteading as a return to the domestic arts to solve many of the ills of the world. For Hayes, homesteaders, those engaging in this radical homemaking, are moving the country away from the extractive economy "where corporate wealth was regarded as the foundation of economic health, where mining our earth's resources and exploiting our international neighbors was accepted as simply the cost of doing business" to a life-serving economy "where our resources are sustained, our waters are kept clean, our air pure, and families meaningful and joyful" (2010, p. 13). Hayes is not alone in seeing homesteading as a radical act. In recent years, a number of books have extolled the virtues of going off the grid, farming, and reclaiming domesticity (Kimball, 2010; Matchar, 2013; Woginrich, 2011; Young, 2011). As such, homesteading is a symbolic movement that firmly roots one in place, in opposition to the global food economy.

A Place-Based Response: The Weston A. Price Foundation

The approach to eating, outlined by the Weston A. Price Foundation (WAPF), gives insight into the reasons for the great importance of place. WAPF was founded in 1999 by Sally Fallon to "disseminate the research of nutrition pioneer Dr. Weston Price, whose studies of isolated nonindustrialized peoples established the parameters of human health and determined the optimum characteristics of humans diets" (Nienhiser, 2000). Dr. Price traveled extensively throughout the world and visited a number of small-scale populations. Upon arrival, he looked into the mouths of every community member to count their dental caries and deformities, noted their general health, recorded their diet, and sent samples of the food they ate to a lab for nutrient and mineral analysis (Price, 2008). He found that hygiene had little effect on dental and bodily health. Rather, diet was the key indicator of overall health. Populations that maintained a diet of unprocessed, locally gathered foodstuffs had few (if any) caries, good general health, symmetrical faces, and good dispositions. Moreover, those individuals in the community who had switched to a modern diet of highly processed foods had higher incidences of dental deformities and health problems. Such maladies were found in their children in alarmingly high rates.

Sally Fallon wrote *Nourishing Traditions* (1999), outlining a healthy diet based on Price's findings. Her thesis is that all traditional societies ate foods that were readily available to them, and this is the principle that current WAPF followers continue to follow. They promote a local, place-based approach to eating. By eating locally, WAPF followers feel they can ensure that the foods they consume are grown and processed in an acceptable manner. Foods that are "traditional," as defined by WAPF, usually do not adhere to the health standards touted by the mainstream nutritional community. Fallon, for example, claims to eat a diet composed of 60 percent fat and recommends slathering toast with a smear of butter as thick as the toast itself. WAPT followers do not support a vegetarian or a vegan diet: rather, a diet comprising many animal products, particularly raw milk, egg yolks, butter, organ meats, grass-fed beef, and pastured pork and poultry. They argue that animal products are nourishing and health-bringing but that food items, which have been subjected to synthetic feed, unnatural living conditions, or any other conditions that defy how they were meant to be raised, are unhealthy and should not be consumed. In many ways, WAPF followers fall into the "local trap" (Born & Purcell, 2006). Local is perceived as being inherently better because of what it is not: large and commercial.

To eat according to WAPF principles, followers must be acutely aware of how their food is made and where it comes from. For example, at a 2013 WAPF conference, the most common topic of conversation between sessions was where to buy products that one would feel safe eating. For example, one participant started her own homestead because it was so difficult to buy raw milk where she lived. Another theme of this conference, evident both in the presentations and in participants' conversations between sessions, was that we need to get back to eating the way of the farm. Only a few generations ago, individual eaters used to source and process their own foods. WAPF followers believe that when this connection to food and place was lost, a myriad of social, psychological, and health problems arose. Reconnecting with the past, with how we used to be, necessitates a reconnection to the land and to place. A quote from homesteader Kiera (a pseudonym) illustrates this point well: "We walk past people, we don't look at them. We get our food at a store, we don't know where it came from...if we just started to look to what the human species is supposed to look like, what we're supposed to do, we would radically change" (personal communication, 2013). Eating this way requires a return to the land.

Weston A. Price's followers use all the strategies mentioned in the previous section as part of their quest to eat in accordance with what they are biologically and, for some, spiritually supposed to consume. They source foods at local farmers markets, buy cow shares for meat and milk, visit farms, and homestead. Gretchen (a pseudonym), a woman very involved in her local WAPF chapter, provides an excellent example of the buying strategies that "WAPF-ers" use. At the time of our interview (2013), she and her family had recently moved to a rural area and lived in a small country house surrounded by several acres of land. They had not yet begun to raise animals or to grow crops, but they were planning on it. In the meantime, they sourced their foods from a variety of farms. They had purchased four raw milk shares and picked up their milk weekly from one farm. Sometimes they would get their eggs from there as well if the neighbors had none to share. They got their fruits and vegetables through a CSA and had to pick their box up at the farm as well since they lived in a rural area far from a drop-off spot. Meat came from yet another farm and yet another pickup.

Not all WAPF followers are as dedicated as Gretchen, but all are very aware of where their food comes from and what the labels attached to them mean. This means a complete reconceptualization of the food landscape in which foods that are manufactured are dangerous. Foods can be labeled as healthy simply as a factor of being produced in a known place. Salatin, who is greatly admired in the community, speaks of honoring the "pigness of the pig" (another oft-cited colloquialism of his) by allowing the pig to live its life as it was meant to be lived. In this way, that pig will live a good life in the pasture and then provide nourishing meat.

An Issue of Resources

It is important to note that as influential as the food movement might be, it is still predominantly a middle- and upper-class movement. While one does not have to be economically rich to eat a local, sustainable diet, done does need to be resource rich. Arguably, the most important of these resources is time. First, there is time involved in scouting out food products. This often takes several trips to farms, farmers markets, pickup locations, and stores, as the example of

Gretchen indicates. Once the food has been purchased, preparation can take copious amounts of time. For example, Gretchen's husband lamented that he could never just grab something to eat. Many WAPF followers note that meal planning is essential. It is essential not just because they need to plan what to buy but also how to prepare foods in a traditional manner (soaking, sprouting, and fermenting) because it can take days.

Another resource is knowledge. This is not just knowing about which foods meet health standards but also where to access knowledge on buying and preparing food. Followers of these approaches speak of a "steep learning curve" but argue that once you know what you are doing, it is not especially time-consuming or considerably more expensive than mainstream approaches to eating. However, this knowledge is often inaccessible. How to source foods and prepare them is not necessarily readily available knowledge. Moreover, nutritional knowledge can be conflicting. It is difficult to know how to evaluate such knowledge. WAPF, for example, rejects a vegetarian/vegan diet, but such a lifestyle has long been promoted as healthy and life-affirming. There is no definitive answer as to which is the correct approach.

Embeddedness is important to consider when highlighting the lack of access to knowledge resources. The connection between the purchaser and the farmer can be utilized to address this gap in knowledge. With their produce, vendors at farmers markets often offer cooking tips or share recipes. They encourage consumers to try new products, and they hawk "ugly fruits and vegetables" (produce that does not meet grocery stores' aesthetic standards). These informative interactions are common throughout markets and help ensure the success both of vendors and of purchasers.

A third important resource is geography. Local eaters need to have access to farms, farmers markets, and/or supportive communities where there are drop-off locations for CSAs. Although growing in number, such resources still do not reach those who are most in need of healthy foods, including people who live in food deserts without access to affordable, healthy food. There have been some initiatives to reach this population, but they have often been limited in their scope because monetary, time, and knowledge resources are limited. I interviewed an individual who started a farmers market in an urban food desert to get fresh foods to low-income people. He built it and was very angry when they did not come. But he had to admit that the food he offered was unfamiliar to the community members and took longer to prepare, even if they were closer and the products were affordable (personal communication, 2013).

Monetary resources are, of course, also vitally important. Local food advocates are fond of mentioning the prices of sodas and fast food to make the argument that a locally prepared meal can be just as cheap, if not cheaper. However, the costs are distributed differently. Those who eat locally often need to buy ahead and in bulk. Meal-to-meal, the outlay of expenses may be the same or even less than conventionally bought and prepared food, but most of the money was paid up front. This is not an option for those who live paycheck to paycheck. However, some farmers markets accept SNAP (Supplemental Nutrition Assistance Program) and WIC (Women, Infants, and Children) benefits, and the number that does so is increasing.

Conclusions and Observations

In this chapter, I have shown how a strong sense of place is often proposed as a solution to the nutritional, environmental, and psychological deficits of the "big food" agribusiness complex. A place-based approach to eating and agriculture, such as the one promoted by WAPF, requires a complete rethinking of our food economy and local landscape. One of the biggest critiques of the local food movement is that it is limited in scope. Thus, it works for some, but large food producers are still needed to feed the masses. In other words, our population is too large to support small-scale agriculture. Understandably, local food activists take issue with this and argue that eating small is possible if we rethink our approach to food. First, more land should be turned over to food production. The trend has been, and continues to be, a reduction of agricultural land. Therefore, more food is produced on smaller amounts of land controlled by fewer people, which leads to the second point. More people should be involved in food production. Arguably, everyone should, in some way, be involved in food production: in a small way, by growing some vegetables and keeping backyard chickens; or in a large way, by raising livestock. Food production should not just be the work of a few farmers but of everyone. Finally, we should eat food that is produce-able. This means that if it cannot be grown, raised, or prepared by the home gardener or cook, it should not be eaten. To eat locally is not just to make an economic or dietary choice but a symbolic and cultural one based on the belief that food is, and must be, intimately tied to place.

References

Allen, W. (2012). *The good food revolution: Growing healthy food, people, and communities.* New York, NY: Gotham Books.

Barber, D. (2014a, May 17). What farm-to-table got wrong. *New York Times Sunday Review*. Retrieved from http://www.nytimes.com/2014/05/18/opinion/sunday/what-farm-to-table-got-wrong.html?hp&rref=opinion&_r=2

Barber, D. (2014b). *The third plate: Field notes on the future of food*. New York, NY: Penguin.

Born, B., & Purcell, M. (2006). Avoiding the local trap: Scale and food systems in planning research. *Journal of Planning and Education Research, 26*(2), 195–207.

Canty, K. (Producer & Director). (2011). *Farmageddon: The unseen war on America's family farms* [Motion picture]. United States: Passion River.

Couric, K., & David, L. (Producers), Soechtig, S. (Director). (2014). *Fed up* [Motion picture]. United States: Atlas Films.

de St. Maurice, G. (2014). The movement to reinvigorate local food culture in Kyoto, Japan. In C. Counihan, & V. Siniscalchi (Eds.), *Food activism: Agency, democracy, and economy* (pp. 77–94). New York, NY: Bloomsbury.

Fallon, S., & Enig, M. G. (1999). *Nourishing traditions: The cookbook that challenges politically correct nutrition and the diet dictocrats*. Washington, DC: New Trends Publishing.

Glowacki-Dudka, M., Murrary, J., & Isaacs, K. P. (2013). Examining social capital within a local food system. *Community Development Journal, 48*(1), 75–88.

Gould, R. K. (2005). *At home in nature: Modern homesteading and spiritual practice in America*. Berkeley: University of California Press.

Hayes, S. (2010). *Radical homemaking: Reclaiming domesticity from a consumer culture*. Richmondsville, NY: Left to Write Press.

Hinrichs, C. C. (2000). Embeddedness and local food systems: Notes on two types of direct agricultural market. *Journal of Rural Studies, 16*(3), 295–303.

Imhoff, D. (Ed.). (2010). The CAFO reader: The tragedy of industrial animal factories. Healdsburg, CA: Watershed Media.

Janssen, B. (2010). Local food, local engagement: Community-supported agriculture in eastern Iowa. *Culture and Agriculture, 32*(1), 4–16.

Kimball, K. (2010). *The dirty life: A memoir of farming, food, and love*. New York, NY: Simon & Schuster.

Kloppenburg, Jr., J., Hendrickson, J., & Stevenson, G. W. (1996). Coming in to the foodshed. *Agriculture and Human Values, 13*(3), 33–42.

Lappé, F. M. (1971). *Diet for a small planet*. New York, NY: Random House.

Lyson, T. A. (2012). *Civic agriculture: Reconnecting farm, food, and community*. Lebanon, NH: University Press of New England.

Matchar, E. (2013). *Homeward bound: Why women are embracing the new domesticity*. New York, NY: Simon & Schuster.

McFadden, S. (2014, April 24). Historic pivot point for food democracy. *The Call of the Land*. Retrieved from thecalloftheland.wordpress.com/2014/04/24/historic-pivot-point-for-food-democracy

Nienhiser, J. C. (2000). About the foundation. The Weston A. Price Foundation. Retrieved from http://www.westonaprice.org/about-the-foundation/about-the-foundation/

Pearlstein, E., & Kenner, R. (Producers), Kenner, R. (Director). (2008). *Food, Inc.* [Motion picture]. United States: Magnolia Pictures.

Planck, N. (2006). *Real food: What to eat and why.* New York, NY: Bloomsbury.

Polanyi, K. (1957). The economy as instituted process. In K. Polanyi, C. M. Arensberg, & H. W. Pearson (Eds.), *Trade and market in the early empires* (243–270). Glenscoe, IL: Free Press.

Pollan, M. (2006). *The omnivore's dilemma: A natural history of four meals.* New York, NY: Penguin.

Pollan, M. (2008). *In defense of food: The myth of nutrition and the pleasures of eating.* New York, NY: Penguin.

Pollan, M. (2009). *Food rules: An eater's manual.* New York, NY: Penguin.

Pollan, M. (2010, June 10). The food movement, rising. *The New York Review of Books.* Retrieved from http://michaelpollan.com/articles-archive/the-food-movement-rising/

Price, W. A. (2008[1939]). *Nutrition and physical degeneration.* Loveland, CA: Price-Pottenger Nutrition Foundation.

Putnam, R. D. (2000). *Bowling alone: The collapse and revival of American community.* New York, NY: Simon & Schuster.

Robbins, J. (2011). *The food revolution: How your diet can help save your life and our world.* San Francisco, CA: Conari.

Salatin, J. (2007). *Everything I want to do is illegal: War stories from the local food front.* Swoope, VA: Polyface.

Salatin, J. (2011). *Folks, this ain't normal: A farmer's advice for happier hens, healthier people, and a better world.* New York, NY: Hachette.

Schlosser, E. (2001). *Fast food nation: The dark side of the all-American meal.* New York, NY: Houghton Mifflin.

Shiva, V. (2000). *Stolen harvest: The hijacking of the global food supply.* Cambridge, MA: South End Press.

Sonnino, R. (2007, March). The power of place: Embeddedness and local food systems in Italy and the UK. *Anthropology of Food.* Retrieved from aof.revues.org/454

Thivet, D. (2014). Peasants' transnational mobilization for food sovereignty in La Vía Campesina. In C. Counihan, & V. Siniscalchi (Eds.), *Food activism: Agency, democracy, and economy* (pp. 193–209). New York, NY: Bloomsbury.

US Department of Agriculture (USDA). (2013). Small farmers & urban agriculturalists: Fact sheet—January 2013. Retrieved from http://www.nrcs.usda.gov/Internet/FSE_DOCUMENTS/stelprdb1083296.pdf

US Department of Agriculture (USDA). (2014). National count of farmers market directory listing graph: 1994–2004. Retrieved from http://www.ams.usda.gov/AMSv1.0/ams.fetchTemplateData.do?template=TemplateS

&leftNav=WholesaleandFarmersMarkets&page=WFMFarmersMarketGrowth&description=Farmers+Market+Growth

Vasavi, A. R. (1994). Hybrid times, hybrid people: Culture and agriculture in south India. *Man, 29*(2), 283–300.

Weiss, B. (2011). Making pigs local: Discerning the sensory character of place. *Cultural Anthropology, 26*(3), 438–461.

Weiss, B. (2012). Configuring the authentic value of real food: Farm-to-fork, snout-to-tail, and local food movements. *American Ethnologist, 39*(3), 614–626.

Whorton, J. C. (1994). Historical development of vegetarianism. *American Journal of Clinical Nutrition, 59*(5), 1103S–1109S.

Woginrich, J. (2011). *Barnheart: The incurable longing for a farm of one's own.* North Adams, MA: Storey Publishing.

Wilk, R. (2006). From wild weeds to artisanal cheese. In R. Wilk (Ed.), *Fast food/slow food: The cultural economy of the global food system* (pp. 13–27). Lanham, MD: AltaMira Press.

Winter, M. (2003). Embeddedness, the food economy and defensive localism. *Journal of Rural Studies, 19*(1), 23–32.

Young, T. (2011). *The accidental farmers: An urban couple, a rural calling and a dream of farming in harmony with nature.* Elberton, GA: Harmony Publishing.

Chapter 11

Troubling Place in Alternative Food Practices: Food Movements, Neoliberalism, and Place

Janna Lafferty

Introduction

The movement for local, organic, sustainable agriculture and food provisioning in the United States is framed as resistance to the expansion of corporate agriculture. In a coherent critique of industrial farming, US alternative movement practitioners often foreground the loss of locality and an ethic of place that is attendant with the Green Revolution, and the globalization of intensive, mechanized agriculture. Emphasizing the environmental costs of this global shift in food production, the mainstream alternative food movement builds on an "agrarian imaginary" (Guthman, 2004) that invokes a more ecologically sustainable agricultural past, which is characterized primarily by small, family farm production. One goal is to bring urban and suburban consumers into more direct relationships with farmers by re-localizing food systems. Celebrity chefs, alternative food gurus, and authors have popularized a set of tastes and consumption habits as a way of creating ecologically and socially conscious practices at food production sites, a trend scholars refer to as *ethical consumption*. Urged to "vote with our forks" by purchasing products labeled Organic, Fair Trade, or Geographically Indicated, the mainstream alternative food movement has been at the forefront of building a moral economy of consumption in the Global North, whereby consumerism constitutes engaged citizenship.

Increasingly, scholars and activists have troubled alternative food-movement maxims, values, and ideals, raising questions about whether alternative food practices resist or replicate a neoliberal capitalist food regime and its constitutive social privilege. Many scholars have highlighted the economic prohibitions in which participation in the movement requires buying power, and others have illustrated how cultural codings and spatial practices have contributed to raced and classed exclusions and erasures of diverse food cultures (Alkon & Agyeman, 2011; Guthman, 2008; Slocum, 2007). Food-systems scholars reveal similar disparities between global food security discourses and the realities they obscure in local places. The international community prioritizes ending global hunger but systematically excludes local communities from policy-making decisions about how to achieve food security (Patel, 2009). Top-down, market-oriented policies in the name of food security, food aid, or sustainable agricultural development frequently fail to consider local objectives or traditional knowledge and consequently result in land loss, cultural assimilation, and dispossession (Patel, 2009).

A proliferation of food-centered social movements challenges the exclusivities of the mainstream alternative food movement and food-security discourses. Researchers have variously conceptualized how this multiplicity of food movements relate and what they can help us understand about cultivating a meaningful politics for transforming the food system. In this chapter, I draw from food studies to complicate ideas of place and food, suggesting that an idealized return to local, small-family farm production and market-based efforts—to protect the ecologies and cultures of places—may be flawed approaches to the social and environmental problems wrought by the current global food system. I highlight the ways in we might think differently about place in relation to food-system change to meaningfully attend to spaces and places that are excluded, not only from access to healthy, culturally appropriate food, but also to democratic inclusion in defining and governing just and sustainable food systems.

Food Movements and the Global Food System

Food movements abound. Food crises, global hunger, food-related chronic disease, and global ecological degradation are widely recognized as problems enmeshed with the status quo of food production, distribution, and consumption. The divergent models advanced for grappling with these issues constitute the contested terrain of food politics, where *politics* is defined in the broadest sense. Thousands of food movements and uprisings confronting these exigent issues speckle the United States and the globe. However, identifying the extent to which they challenge what scholars refer to as a global food system, or corporate food regime, it is possible to make sense of this terrain and of the kinds of futures that various food movements point toward. Eric Holt-Giménez and Annie Shattuck identify a fourfold typology that encapsulates the ways global political food trends broadly relate to the current corporate food regime. Neoliberal and reformist trends ultimately preserve the global food system, while progressive and radical trends challenge it (2011).

The notion of a global food system came out of sociologist Harriet Friedmann's work in the 1980s. She identified "a rule-governed structure of production and consumption of food on a world scale" that is central to the emergence and reproduction of global capitalism (Friedmann, 1993, qtd. in McMichael, 2009). Her insight generated a field of scholarship, dedicated to food systems analysis, which examines the institutions, ideas, and policies that govern how food is produced, distributed, and consumed, and the social relationships they structure at various social scales (Guthman, 2008). Although scholars debate whether the regime Friedmann described is better characterized as a system or a network, it is largely acknowledged that certain countries, institutions, and actors wield far greater decision-making power than others in determining the socioenvironmental structures it takes. The dynamics defining that power structure come out of colonial-era relationships between empires and their colonies as well as from settler states and colonized populations.

The global food regime first emerged through European exploitation of raw materials and labor from its tropical colonies to fuel industrialization, beginning in the late nineteenth century (Holt-Giménez & Shattuck, 2011; Mintz, 1985). Anthropologist Sidney Mintz most famously illustrated how the intensification of sugar consumption, sustained by enslaved African labor in the Caribbean, provided the cheap calories that sustained an emerging working class in England, thereby fuelling the Industrial Revolution (1985). While the transatlantic sugar trade predated capitalism, Mintz notes that it forged a kind of proto-industrial capitalism. Sugar plantations needed to be mechanistically choreographed in ways that anticipated the regimentation of industrial factories. They required capital investment and reinvestment to produce profits and accumulate wealth. Plantations engendered core-periphery metropolitan economies through processes of accumulation by dispossession and necessitated the market expansion. During this period, sugar, wheat, and meat from European colonies in the Western hemisphere supplied Europe with requisite food staples to support a proletarian class (Holt-Giménez & Shattuck, 2011; Friedmann & McMichael, 1989).

The mid-twentieth century marks a reversal of flows between the Global North and South, in which American politicians and powerful philanthropic foundations secured geopolitical desires by pushing for industrial agricultural infrastructure in the so-called Third World after the Second World War (Cullather, 2010; Holt-Giménez & Shattuck, 2011). By "dumping" agricultural surpluses in Southern countries under the banner of food aid, the United States created dependencies upon US-led agricultural markets. Subsequently hailed as the *Green Revolution*, the implantation of industrial agriculture in the Global South weakened peasant food economies, giving large landowners and capital investors increased control over productive land. Not only were Green Revolution projects socially disruptive, as they materialized in countries in the South, but they also often failed on their own terms. Irrigation in Afghanistan, for example, created soggy, unusable land (Cullather, 2010). As land became privatized or taken over by capital-intensive agricultural operations, subsistence farmers were displaced en masse from productive lands onto infertile hillsides or urban slums (Holt-Giménez & Shattuck, 2011). While yields increased dramatically, they did not lead to fewer hungry stomachs. People starved, not because of food shortages, but because their poverty restricted access to the abundance that others enjoyed. Although the Green Revolution appears successful in that it produced enormous surges in production by rationalizing and industrializing agricultural production, increasingly, scholars and activists highlight that to date, technological increases in yields have done very little to mitigate rural poverty or chronic starvation across the globe (Cullather, 2010).

Whether we have entered a third stage in the global food regime, or global capitalism more generally, remains a matter of scholarly debate. While some scholars question if the term *neoliberalism* offers conceptual purchasing power or represents distinctly novel processes, most social scientists invoke neoliberalism to characterize trends in global political economy and changes in governance since the 1990s. For many scholars, neoliberalism represents "a shift from the state as a regulatory force to a facilitator of markets" in which privatization and entrepreneurialism become agents of social welfare and entitlement (Alkon, 2013). Many social scientists stress the precarity of that shift, which converts government-protected rights into commodities, whereby individual consumers define and address social and environmental problems in the private sphere. Whether or not neoliberalism offers a coherent conceptualization of political economic and ideological shifts, the trends that scholars have used the term to describe are widely acknowledged: breakdowns in barriers to international flows of capital; accelerated privatization of public resources; hyperindividualized ideas about health; the retreat of the state and influx of NGOs (nongovernmental organizations) and private actors as managers of social welfare. These shifts function, at base, to meet capitalism's need for endless expansion (Harvey, 2005).

Food scholars Eric Holt-Giménez and Annie Shattuck assert that the 1980s mark a transition into a neoliberal "corporate food regime," marked by tariff removals, structural adjustment programs, diminishing national marketing boards, and, ultimately, unprecedented market power and corporate monopolization over food production (2011). During this period, the World Trade Organization (WTO) facilitated the liberalization of agricultural trade, mainly by establishing international trade rules that forbade nations in the Global South from regulating agricultural trade in ways that emerging settler states enjoyed when they were developing economically. Thus for these countries, national food security is not only contingent upon the vulnerabilities of climate, but also by undulations in the global market.

Neoliberalism and the Alternative Food Movement

In a 2006 Op-Ed, Michael Pollan summarized the growing tide for local and organic food as "market as movement" (2006). His exhortation to achieve social change through the market via consumer practices instantiates a neoliberal philosophy asserting "the primacy of the market in attending to human needs and well-being" (Alkon & Mares, 2012, p. 348). Pollan's vision, and the alternative food movement he has helped cultivate, generally circumvents traditional forms of political engagement in favor of "voting" at the checkout counter. Ethical consumption as a proxy for political engagement has become not only a dominant mode of social and environmental engagement, but also a thoroughgoing identity construction, particularly in metropolises in the Global North. In her multisited ethnographic research on coffee as a global commodity, anthropologist Paige West discovers that consumers in New York, Sydney, and London coffee shops rehearse identity scripts about choosing products sourced through fair-trade operations (2012). Consumers express affective and therapeutic sentiments about patronizing brands marketed as environmentally and social responsible and the sense that they are helping to alleviate problems such as poverty and climate change through their purchasing choices. Yet her work among coffee producers in Papua New Guinea and with distributors along the commodity chain reveal that the rhetoric and images marketed by those businesses do very little to improve the welfare of coffee producers. When pressed, consumers in cosmopolitan coffee shops knew little or had incorrect information about what fair trade and organic labels mean in practice, or the cultural or environmental conditions of coffee producers in Papua New Guinea, from where their coffee was sourced.

Certification ostensibly offers beneficial services all around: protecting the environment, educating poor people about how to have more efficient and lucrative farms, increasing shareholder value, depending less on chemical inputs, protecting endangered species and ecosystems, and ensuring better product quality. It is supposed to set up more equitable and direct relationships between workers, traders, and consumers. But in this narrative, the producers and consumers described are the fantastical products of careful marketeering efforts. Ironically, their marketing strategies rely precisely on images of poverty and primitiveness to create added value. Anthropologist Sarah Besky points to similar dynamics at work in attempts to bring tea plantations in Darjeeling, India, under the auspice of Fair Trade. She draws from scholarship on "hyperreality," describing consumer culture's increasing proclivity for simulacra, in which consumers prefer "fake" experiences like Disneyworld's EPCOT "world showcase" (Besky, 2013, p. 112). Fair trade renders Darjeeling plantations "both productive spaces and 'simulations'—performances in which the reality of agricultural labor and perceptions of it stand in tension" (2013, p. 112). She illustrates this by revealing how fair trade and other neoliberal schemes for justice, such as microloans, stand in tension with Darjeeling tea plantation workers' visions for justice.

In one vivid anecdote, Besky recounts how a Nepalese tea plantation worker dealt with the cow that was supplied to her through a microloan intended to supplement her meager work wage. Selling the cow's milk for additional income, as the fair trade company intended, was not realistic for the woman, who not only worked such long days that she did not want to put in extra hours milking a cow, but could not stomach selling milk to her neighbors for her individual profit. Instead, she let her neighbors have free use of the cow—commonsense in the context of the local moral economy. Besky drives home the point that the woman's cow "represents the individualizing tendency of fair trade," which challenges "long-standing moral economic relationships between Darjeeling tea-plantation laborers, land, management, and the postcolonial state" (Besky, 2013, p. 114). As a neoliberal project that aims to enact social justice through market mechanisms, fair trade merely undermines existing state structures safeguarding equitable and humane treatment. Exchanging state enforcement of plantation owners' obligations to provide services for workers with voluntary market schemes only leaves plantation workers more vulnerable. "Fair trade's vision [...] accounts for neither long-standing moral economic relationships within plantations nor institutions already in place to support plantation labor" (2013, p. 114–115). It is, rather, those in positions of wealth and power who benefit from fair trade: international fair trade and organic certifiers, plantation owners, and the tea retailers who get to market these labels to increasingly "ethically conscious" consumers. The Darjeeling tea plantation workers with whom Besky interacted unequivocally expressed their desire for the enforcement of Indian labor laws over the version of social and environmental responsibility enacted by market schemes.

West's and Besky's work illustrate how Western imaginaries of "the primitive" are exploited, naturalized, and Othered in ways that make programs with slightly less offensive practices seem philanthropic, because poverty appears to be a kind of natural condition or starting point for them. The neoliberal subjectivities produced through ethical consumption do not meet the visions of justice imagined by food producers, or by the local people targeted as beneficiaries of this practice. While fair-trade discourses claim to offer an alternative to the food system status quo, fair trade ends up reproducing a neoliberal global political economy of food in which markets and individual consumers constitute the primary arena for meeting social and environmental needs. Julie Guthman's work on organic farming in California similarly illustrates how putatively "alternative" food practices end up replicating industrial agriculture, which they set out to change. She finds that the very industrial processes its proponents claim to challenge have absorbed organic agriculture in California—the United States's largest organic farming sector. Organic produce often comes from small plots on big industrial farms that simply use less chemical inputs in those areas (Guthman, 2004).

Yet organic food and farming evoke a powerful "agrarian imaginary" in the United States that sustains the value of this label (Guthman, 2004). Guthman's notion of the "agrarian imaginary" points to the trafficking of images in the United States of a pristine and righteous agricultural past that can be recovered. It offers a redemptive narrative achievable especially through a return to family farm production. Ideas about the righteousness of working the land and touching the soil has its roots in colonial American discourse (Besky, 2013), obscuring stark histories of racial exclusion in US agricultural history. The United States has, since it origins, prevented landownership through racial discourses that persist across every level of industrial agriculture today (Slocum, 2010). American expansion was achieved partly through land giveaways that not only appropriated the lands of Indigenous nations, but also disenfranchised Hispanos and Californos in the Southwest, denied African Americans ownership in the South, and precluded Chinese and Japanese immigrants from landownership (Guthman, 2012). Land giveaways were proffered exclusively to whites, extending and solidifying "the racialized land-labor relationships that had begun with slavery," in which whites owned the land and others labored it by way of compulsion or economic desperation (Guthman, 2012). The 1865 Homestead Act essentially handed out land to white farmers in the West to achieve continental settlement. Today, most of the organic produce put on market shelves in the United States is not the product of the small family farm pictured by the agrarian imaginary, but of vaguely sequestered sections of large industrial farms worked by racially marginalized farmworkers (Guthman, 2004).

Guthman goes on to argue that the alternatives presented by the alternative food movement create places and people that cannot be served by these alternatives, such that some places and people are essentially ignored (2012). She observes that "alternative food movement gurus offer a fairly limited menu of solutions to a deeply problematic food system, reflecting cultural values and social power of those who fare well under contemporary capitalism and invested in particular ways of seeing the problem" (2011, p. 17). Alison Hope Alkon and Julian Agyeman have observed that the US alternative food movement is overwhelmingly constituted and curated by white, middle-class people with similar values and backgrounds, itself constituting a kind of "monoculture" (Alkon & Ageyman, 2011). Whites who are invested in this movement "often simply do not see the subtle exclusivities that are woven into its narrative" (Alkon & Agyeman, 2011, p. 3). Considerations of the roles that race, class, and ethnicity play in food politics and the food system and diversity in the meanings of food are largely neglected in alternative-food discourses. Yet they play a profound role in shaping the food system itself, as well as in determining whether one can participate in the kinds of consumption strategies that the mainstream food movement proposes will change that system (Alkon & Ageyman, 2011; Guthman, 2012).

In her research, Clare Hinrichs finds that "many direct agricultural markets involve social relations where the balance of power and privilege ultimately rests with well-to-do consumers" (2000, p. 301). The transparency of whiteness renders these exclusions and erasures invisible and unproblematic; the alternative food movement ostensibly reaches out to anyone, who only needs to make the "choice" or set of choices to eat healthfully and to support sustainable agriculture. Food scholars have begun to question whether getting vegetables into inner cities and telling people how to eat healthfully does very much to address the broader political economic inequities that shape racialized disparities in food insecurity, chronic disease, and poverty (Guthman, 2008).

Holt-Giménez and Shattuck have outlined the differences between what they see as, on the one hand, neoliberal and reformist programs and, on the other, progressive and radical movements for addressing food-related social and environmental problems. Among those they identify as representing a "reformist" trend are included "the corporate mainstreaming faction of Fair Trade," those promoting "'responsible' foreign direct investment in agricultural land [. . .], the various industry-dominated 'roundtables' for sustainable soy, palm oil and biofuels; corporate sectors of the organic foods industry; and civil society driven corporate social responsibility and industry self-regulation initiatives" (2011, p. 122). They also point out that many humanitarian and environmental organizations "are wholly or partially rooted in the Reformist trend, partly because their main sources of funding come from government, major corporations, or large philanthro-capitalist institutions such as the Bill and Melinda Gates Foundation" (2011, p. 122). They specifically name Bread for the World, Oxfam-America, CARE, WorldWatch, World Vision, and International Federation of Agricultural Producers (IFAP) as exemplary of this trend. Seeking to incorporate less environmentally and socially degrading alternatives into existing market structures, these actors promote incentive-based certification projects and corporate self-regulation, approaches that aim to "modify industrial behavior through the power of persuasion and consumer choice" (2011, p. 121).

They explain that the underlying notion here "is that by dint of a good example or 'voting with our forks,' less damaging trade and production alternatives will someday transcend their market niches (frequently high-end specialty products and set new industrial standards" (2011, p. 121). What actually emerges, they argue, "is an uneasy dualism between 'quality food' for higher income consumers and 'other food' consumed by the masses" (2011, p. 121). The reformist model, then, is basically compatible with capitalist overproduction, which proliferates cheap, unhealthy food, relies on unstable yield-intensification technologies with harmful long-term social and ecological effects—such as hybridization, chemical pesticides, and herbicides, and captures farmers in a technology treadmill characterized by falling prices and the ongoing need for investments in new technologies to stay afloat (Guthman, 2012). Without challenging extant capitalist structures, it merely calls for things like third-party certification programs to address flagrant sustainability and social justice problems: fair trade, organics, or voluntary industry roundtables (Holt- Giménez & Shattuck, 2011).

Neoliberalism fetishizes the market by describing it as working on its own, rather than as comprising social relations and human labor. Ethical consumerism implies that individuals are, or can be, responsible for conditions of production, including social justice and environmental impacts. In this narrative, the governments and institutions that established the structural adjustment programs and agricultural trade policies that have engendered or exacerbated poverty, to begin with, are never acknowledged or held accountable. The inherent problems and contradictions of global capitalism remain unquestioned, and the crises produced by the global commodification of food are met with more market-driven solutions. In reproducing a neoliberal political economy, so too are racialized geographies of privilege. The alternative food movement has done little to call into question the invisibility of socially marginalized food producers inside and outside the United States, instead romanticizing the image of colonial and Western expansion–era landholders as the ur-organic farmer. This image leaves obscured cultural and geographic constraints to accessing organic produce and farmers markets, reproducing assemblages of mostly white bodies in "alternative" food spaces (Slocum, 2010).

Social-Justice Oriented Food Movements

Food Sovereignty

As part of a transnational peasant coalition called La Vía Campesina, Indigenous peoples, peasants, smallholder farmers, and fisherfolk invented the notion of *food sovereignty* as a counter-discourse to the "top-down" antihunger narratives of international development actors (Patel, 2009). It is a product of negotiation and imagination, as well as a cognizance of how the global food system engenders and accelerates interlinked social and ecological crises worldwide (Desmarais & Wittman, 2014). As this framework is espoused by diversely situated communities across the globe, emplaced food-sovereignty struggles take on different inflections in relation to the historical, ecological, political, and cultural contests in which they are embedded. Particularly in the Americas—but also in Africa, Asia, and Europe—food sovereignty is being integrated with struggles around Indigenous identity vis-à-vis self-governance, territory, and environmental use. Differences—even contradictions—exist in how food sovereignty is articulated in diverse contexts (Patel, 2009; Desmarais & Wittman, 2014). Yet they share many basic goals as well as a sense of how their local struggles are embedded in broader processes of global capitalism, reflecting what Simon Springer calls "relational geographies of resistance" that acknowledge the local and global to be mutually constituting (2011). Food sovereignty activists promote the relocalization of food systems, but unlike the mainstream alternative-food movement, they emphasize empowerment of local communities by implementing structures that guarantee more control over food systems, which requires directly challenging a neoliberal political economy of market rule (Alkon & Mares, 2012).

As an organizing frame for comprehensive social change, food sovereignty involves a shifting set of diverse struggles that are imagined as interlinked. Food-sovereignty struggles tend to understand the problems they address as arising from a global food system, but also as "rooted in *local* and *national* struggles of dispossession" and destruction of livelihoods (Bush, 2010, p. 121; Desmarais & Wittman, 2014). Hunger is not understood as a problem of "low productivity, unemployment, poor wages or inadequate distribution, but by inequities in the determinants of production, reproduction and distribution, [or] the entitlements extending to relations of exchange, modes of production, social security and employment" (Holt-Giménez & Shattuck, 2011). Food-sovereignty activists, then, tend to consider local/organic food movement tactics "insufficient to mount a systemic critique of corporate agriculture and liberal capitalist economics as a whole" (Magdoff, Foster, & Buttel, 2000, p. 188). In this framework "justice" is not conceived merely as increased pay, resource access, or even "the right to food"—all of which leave untouched the question of who controls the food system but, rather, the "right of peoples to define their own food, agriculture, livestock and fisheries systems" (Patel, 2009). Food-sovereignty communities envision fundamental social transformation through food and agriculture (Desmarais & Wittman, 2014). Desmarais and Wittman suggest that "engaging with the concept of food sovereignty as it has evolved among grassroots actors requires a critical engagement with a new politics of possibility" (2014, p. 15). It involves "reshaping the political spaces in which decisions and values shift concerning issues related to how and what food is produced, accessed and consumed" (Desmarais & Wittman, 2014). In their research on how "foodies, farmers and First Nations" take up the language of food sovereignty in Canada, Desmarais and Wittman note how Indigenous food-sovereignty organizations complicate a "local food" ethic and instead put decolonization at the center of an alternative food imaginary. Concerned less with building the local, agriculture-centric food system articulated by the mainstream US alternative-food movement and Canadian "foodies," organizations such as the British Columbia Food Systems Network Working Group on Indigenous Food Sovereignty and the Food Secure Canada Indigenous Circle in Canada "seek to honor, value and protect traditional food practices and networks in the face of ongoing pressures of colonization" (Desmarais & Wittman, 2014, p. 13). Traditional indigenous food-trading networks greatly transcend the locavore "100-mile diet" but function as vital sites of intercultural exchange (Turner & Loewen, 1998). Fishing and hunting are central aspects of self-determination and self-governance. Creating sustainable food systems has less to do with creating areas of conservation and more to do with maintaining traditional food economies, trading networks, and multispecies communities.

Leaders advancing an Indigenous food-sovereignty frame emphasize that colonization and unresolved treaty processes are to blame for extensive territorial and environmental losses that sustain food economies constituted by fishing, hunting, gathering, cultivation, and vast trade networks (Turner & Loewen, 1998; Morrison, 2011; Desmarais & Wittman, 2014). They highlight the direct linkages between these ongoing losses and bodily, cultural, and community well-being (Desmarais & Wittman, 2014, p. 12). At issue are pronounced disparities in rates and experiences of diet-related chronic illness and food insecurity. For Canadian Indigenous food sovereignty leaders, self-determination involves the "freedom and ability to respond to our own needs for healthy, culturally-adapted indigenous foods. It represents the freedom and ability to make decisions over the amount and quality of food we hunt, fish, gather, grow and eat" (Morrison, 2011, p. 100). It specifically identifies the social, cultural, and economic relationships of intertribal and intercommunity food sharing and trading as an apparatus for Indigenous health.

Among the global food sovereignty community—which is organized by international forums such as La Vía Campesina, Nyéléni International Forum for Food Sovereignty, and a multitude of constituent organizations, North American Indigenous food sovereignty leaders have been unique in articulating the sanctity of food and the community of life it entails. The Indigenous Circle of Food Secure Canada emphasizes that "food sovereignty understands food as sacred, part of the web of relationships with the natural world that define culture and community" (People's Food Policy Project, 2011). The Indigenous Food Systems Network articulates that "food is a gift from the Creator; in this respect the right to food is sacred and cannot be constrained or recalled by colonial laws, policies and institutions. Indigenous food sovereignty is fundamentally achieved by upholding our sacred responsibility to nurture healthy, interdependent relationships with the land, plants and animals that provide us with our food" (Indigenous Food Sovereignty Network).

These assertions introduce a set of values and an emplaced moral economy of food that is not acknowledged by dominant alternative food movements or international development bodies working to creating global food security. What is being articulated by these leaders echoes what Sarah Besky identified as a "tripartite moral economy" among Darjeeling tea plantation workers in India, involving not only human social obligations—as elaborated by E. P. Thompson and James Scott—but also obligations between humans and the agro-environment (Besky, 2013, p. 32). Their moral economic understandings of the plantation remained "central to the culture of tea production and to the ways in which tea workers envision its future" in ways that resisted notions of "justice" imposed by fair-trade certifiers, tea marketers, and politicians (2013). So too are moral economic understandings that are central to different Indigenous communities in the North American Pacific Northwest elided in others' projects for food fairness, environmental health, and environmental conservation. As it is for Darjeeling tea laborers, these moral economies involve social obligations between humans and nonhumans.

Food Justice

Drawing from environmental justice discourse, Alison Hope Alkon and Julian Agyeman describe food justice as an extension of environmental justice. Whereas the latter "is primarily concerned with preventing disproportionate exposure to toxic environmental burdens, the food justice movement works to ensure equal access to the environmental benefit of healthy food" (2011, p. 8). The food-justice literature highlights the structural processes through which low-income communities and communities of color have been denied access to healthy food. Where the mainstream alternative-food movement proposes educating ostensibly ignorant people about healthy eating or getting vegetables into urban areas to address issues of access, food justice activists examine how these communities "have been subject to laws and policies that have taken away their ability to own and manage land for food production [while] members of these communities continue to be exploited as farm laborers" (Alkon & Agyeman, 2011, p. 4). Food justice activists are creating and revitalizing self-determined ways of provisioning food for their communities, while rendering visible the racial projects and inequalities inherent in the food system. Food justice imagines a food system in which social justice, public health, and environmental sustainability are prioritized over the industrial agribusiness profits (Alkon & Agyeman, 2011).

Food-justice activism highlights the relationships between food and lived experiences of human categories of difference. Especially where scholars working through a food-justice framework have elaborated processes of what Omi and Winant identified as *racial formation*, some of this work has noted the ways racial difference is coproduced with ecological exploitation (Norgaard, Reed, & Van Horn, 2011).

Sociologists Kari Marie Norgaard and Ron Reed, traditional Karuk dipnet fisherman and cultural biologist for the tribe, have drawn from the food-justice framework to think through the contemporary and historical circumstances that have produced Karuk hunger in Northern California (2011). They describe how a series of "racial projects" damaged Kurak management practices and transferred environmental wealth to non-Indians: genocide, nonrecognition of land occupancy, and assimilation (2011, p. 25). They link these processes to Lisa Sun-Hee Park and David Pellow's broader observation that "racial formation in the United States has always been characterized by an underlying link between ecological and racial domination" (Norgaard et al., 2011, p. 27). In addition to the overt forced assimilation of Indian Boarding Schools, Norgaard and Reed want to highlight the ways in which Karuk cultural food management and production have been made illegal and the forced dependency upon government commodity foods due to depletion of traditional food sources (2011, p. 37).

Currently, state regulations impinge on fishing, burning, hunting, mushroom gathering, and basket-material gathering. They highlight that four dams erected along the Klamath River since 1962 have produced the most dramatic shifts in Karuk diets, since they block access to 90 percent of Spring Chinook salmon-spawning habitat, decimating Chinook populations that fed the Karuk. Along with Alison Hope Alkon and Julian Agyeman, Norgaard and Reed ultimately argue that these efforts exemplify how "communities of color are weaving together new narratives and foodways to create just and sustainable" food systems in which "issues of meaning and identity formation are paramount" (Alkon & Agyeman, 2011, p. 15). Norgaard and Reed carefully point to "how racial and economic inequalities have shaped both the food system and the food movement" in ways that are neglected by dominant narratives for creating a better food system (Alkon & Agyeman, 2011, p. 15). The food justice framework goes beyond alternative food's "agrarian imaginary" and attendant set of practices, which has sought to address food-access issues by teaching people to eat and produce organic fruits and vegetables. Bringing a food-justice lens highlights more structural questions about the production of racialized urban geographies of food insecurity and attends to the importance of culturally diverse food meanings.

Food Workers' Movements

Food-worker exploitation is a marginal issue in mainstream alternative food discourse. Yet those organizing around issues faced by food workers attempt to connect with an array of activisms, including human rights, local/organic agriculture, and food justice. Restaurant and farm workers face some of the most egregious labor abuses, forms of exploitation, and inequalities in the formal economy: the lowest wages, lack of benefits, wage theft, abuse, inhumane conditions, high rates of exposure to toxins, and otherwise physically crippling work. In his "embodied ethnography" of migrant farmworkers in Washington and California, medical and cultural anthropologist Seth Holmes discovers the suffering, racism, and heath disparities entailed in the production of fresh fruit for American consumers (2013). More pointedly, he illustrates how this kind of suffering is naturalized and rendered invisible through attributions of racial difference and citizenship status. Since the Green Revolution increasingly put subsistence farmers out of business in Mexico and the Philippines, landowners in the United States have recruited migrant groups as agricultural laborers who are willing to work for extremely low wages (Guthman, 2012). These ongoing land-labor relationships in the United States remain largely invisible to most American consumers; and the agrarian imaginary harks back to an agricultural past in which the white farmer is a lauded figure, whereas the farmworkers of color who overwhelmingly put food on the plates of American dinner tables remain invisible.

The recent Ram Truck advertisement that aired during the 2013 Super Bowl exemplifies this whitened ideal of farming, as some farmworkers' movements observed. In the days following the Super Bowl, the Coalition of Immokalee Workers (CIW) released a parody of the Ram Truck advertisement, presenting images of the largely nonwhite farmworkers, who were all but missing in the original ad, but who provide the lion's share of agricultural work in the United States. It presented a counterdiscourse to whitened images of American farming and agriculture that was complicit in the perpetuation of farmworker abuses in the United States. CIW provides one of the most successful and visible examples of food workers' movements in the United States. The group began to organize in 1993, finding churches or other small meeting places to deliberate on ways to improve their living and working conditions. In 2001, they advanced the "Campaign for Fair Food" to confront farm-labor exploitation (CIW, 2012). By creating visibility and galvanizing public pressure, CIW pressures multibillion-dollar food retailers into contracts guaranteeing "higher wages, a code of conduct for more humane labor standards and a cooperative complaint resolution system" (Alkon, 2013, p. 10). They subsequently developed an Anti-Slavery Campaign, which brought government officials into cooperation to investigate and emancipate more than 1,200 workers, thus having "pioneered the worker-centered approach to slavery prosecution" (CIW qtd. in Alkon, 2013, p. 10). CIW methods combine traditional political tactics of social movements with innovative ways of leveraging public support to place demands on food and agribusiness. CIW has used hunger strikes, marches, protests, and coalition-building to achieve a number of substantive and important changes.

Food justice, food sovereignty, and food-worker activists are sharply redirecting what ethical practice entails in relation to food, creating a new language to imagine food-system change. These grassroots movements and "organic intellectuals" revisit the past—as imagined by a largely white and socially privileged alternative food movement in North America—to grapple with obscured histories of colonization, dispossession, and exclusion, and their ongoing effects as they relate to food provisioning, eating, and the communities at stake. The "agrarian imaginary" and other white cultural discourses do not work to hold open spaces for diverse alternative food economies, values, and practices, but reinstate a political economy and geographies of social privilege. Rather than shopping one's way into ethical, social, and environmental relationships, food sovereignty, food justice, and food workers' movements engender new obligations to listen to the stories of others (Alkon & Agyeman, 2011,

p. 15), generate interracial coalitions, make room for diversity, and rethink food, not just as commodities people have rights to, but as a set of obligations to trans-species communities of life. Those who are variously enacting food sovereignty, food justice, and food workers' rights are engaged in reconfiguring political spaces of decision-making with regard to food.

Conclusions and Observations

In this chapter, I have reviewed food-studies scholarship that complicates the alternative food movement's relocalization and consumer-driven politics. By introducing several important "alternatives to the alternative" food movement (Guthman, 2008), I have introduced different ways of conceptualizing matters of place and space to address the environmental and social inequities of the global food system.

One of the major critiques of alternative-food practices, underpinned by an "agrarian imaginary," is that these are overwhelmingly white foodspaces due to unacknowledged cultural codings and economic structures of privilege (Alkon & Agyeman, 2011; Hinrichs, 2000; Guthman, 2008; Slocum, 2010). Food-security interventions in underserved areas that are based on alternative food practices such as urban-based community supported agriculture and gardening projects—what Guthman has refers to as "brining good food to Others"—do not seem to meaningfully resonate with targeted beneficiaries and fail to address structural issues that have shaped urban food deserts (2008). The food-justice framework inserts a cognizance of bigger structures that shape how certain spaces, which are often highly racialized, lack access to healthy food. Food sovereignty asks not just how local communities might increase access to healthy, culturally appropriate food, but also how they can have social and political control over their own food systems. Food workers' movements highlight how mobilities and immobilities that are propelled by the emergence and expansion of a global food system have rendered certain bodies invisible, and asks food activists, consumers, and political actors to see past the whitened "agrarian imaginary" of the mainstream alternative-food movement so as to ensure food producers have basic rights—including healthcare, freedom from slavery and exploitation, protection from toxic burdens, and a living wage. Each of these in some way puts capitalism on the table as something to be regulated or altered, in contrast to the mainstream alternative-food movement, which has accepted the market as a major arena through which to enact food-system change.

The sensibilities and concerns that underpin alternative food praxis and ethical consumption are not unimportant, and its accomplishments are not entirely in vain. Holt-Giménez and Shattuck assert that "while food movements have been largely ineffective in ushering in substantive reforms," they "have also given social movements an opportunity to grow, spread, and occupy key political spaces in global and local institutions (2011). They suggest that if advocates of what have proven to be a weak counterdiscourse to the corporate food regime ally with radical movements, substantive transformations in the global system are possible. Conversely, if the more mainstream alternative-food activists lean toward neoliberal agendas, the corporate food system will be fortified. Coalitions for food-system change must be built, but to achieve radically democratic transformation in the food system, they must be oriented toward reembedding markets in societies (Polanyi, 1957), do more to question how capitalism shapes unjust food spaces, and rethink the role of market and consumer strategies in building a place-based food systems ethic.

References

Alkon, A. H., & Mares, T. M. (2012). Food sovereignty in US food movements: Radical visions and neoliberal constraints. *Agriculture and Human Values, 29*(3), 347–359.

Alkon, A. H. (2013). Food justice, food sovereignty and the challenge of neoliberalism. *Food sovereignty: A critical dialogue.* International Conference Paper Series. Retrieved from http://www.yale.edu/agrarianstudies/foodsovereignty/pprs/38_Alkon_2013.pdf

Alkon, A. H., & Agyeman, J. (2011). *Cultivating food justice: Race, class, and sustainability.* Cambridge, MA: MIT Press.

Besky, S. (2013). *The Darjeeling distinction: Labor and justice on fair-trade tea plantations in India.* Berkeley: University of California Press.

Bush, R. (2010). Food riots: Poverty, power and protest. *Journal of Agrarian Change, 10*(1), 119–129.

Coalition of Immokalee Workers (CIW). (2012). About the Coalition of Immokalee Workers. Retrieved from http://ciw-online.org/about/

Cullather, N. (2010). *The hungry world: America's Cold War battle against poverty in Asia.* Cambridge, MA: Harvard University Press.

Desmarais, A. A., & Wittman, H. (2014). Farmers, foodies and First Nations: Getting to food sovereignty in Canada. *Journal of Peasant Studies, 41*(6), 1153–1173.

Friedmann, H., & McMichael, P. (1989). Agriculture and the state system: The rise and fall of national agricultures, 1870 to the present. *Sociologia Ruralis, 29*(2), 93–117.

Garrett, S. M. (2008) 'Nutritional apartheid': bringing justice and groceries back into the 'hood. *Annual Meeting of the Association of American Geographers*, Boston, MA.

Gottlieb, R., & Joshi, A. (2010). *Food justice*. Cambridge, MA: MIT Press.

Guthman, J. (2004). *Agrarian dreams: The paradox of organic farming in California*. Berkeley: University of California Press.

Guthman, J. (2008). Bringing good food to others: Investigating the subjects of alternative food practice. *Cultural Geographies, 15*(4), 431–447.

Guthman, J. (2012). *Weighing in: Obesity, food justice, and the limits of capitalism*. Berkeley: University of California Press.

Harvey, D. (2005). *A brief history of neoliberalism*. Oxford: Oxford University Press.

Hinrichs, C. C. (2000). Embeddedness and local food systems: Notes on two types of direct agricultural market. *Journal of Rural Studies, 16*(3), 295–303.

Holt-Giménez, E., & Shattuck, A. (2011). Food crises, food regimes and food movements: Rumblings of reform or tides of transformation? *Journal of Peasant Studies, 38*(1), 109–144.

Kneen, C. (2011). Food secure Canada: Where agriculture, environment, health, food and justice intersect. In H. Wittman, A. A. Desmarais, & N. Wiebe (Eds.), *Food sovereignty in Canada: Creating just and sustainable food systems* (pp. 80–96). Halifax: Fernwood Publishing.

Magdoff, F., Foster, J. B., & Buttel, F. H. (2000). *Hungry for profit: The agribusiness threat to farmers, food, and the environment*. New York, NY: Monthly Review Press.

McMichael, P. D. (2009). A food regime analysis of the "world food crisis." *Agriculture and Human Values, 26*(4), 281–295.

Mintz, S. W. (1985). *Sweetness and power: The place of sugar in modern history*. New York, NY: Viking.

Morrison, D. (2011). Indigenous food sovereignty—A model for social learning. In H. Wittman, A. A. Desmarais, & N. Wiebe (Eds.), *Food sovereignty in Canada: Creating just and sustainable food systems* (pp. 97–113). Halifax: Fernwood Publishing.

Norgaard, K. M., Reed, R., & Van Horn, C. (2011). A continuing legacy: Institutional racism, hunger, and nutritional justice on the Klamath. In A. H. Alkon & J. Agyeman (Eds.), *Cultivating food justice: Race, class, and sustainability* (pp. 23–46). Cambridge, MA: MIT Press.

Patel, R. (2009). What does food sovereignty look like? *Journal of Peasant Studies, 36*(3), 663–706.

People's Food Policy Project (PPFP). (2011). Resetting the table: A people's food policy for Canada. Retrieved from http://foodsecurecanada.org/sites/default/files/fsc-resetting2012-8half11-lowres-en.pdf

Polanyi, K. (1957). *The great transformation*. Boston, MA: Beacon Press.

Pollan, M. (2006, May 7). Voting with your fork. *New York Times*. Retrieved from http://pollan.blogs.nytimes.com/2006/05/07/voting-with-your-fork/

Slocum, R. (2007). Whiteness, space and alternative food practice. *Geoforum, 38*(3), 520–533.

Slocum, R. (2010). Race in the study of food. *Progress in Human Geography, 35*(3), 303–327.

Turner, N. J., & Loewen, D. C. (1998). The original "free trade": Exchange of botanical products and associated plant knowledge in Northwestern North America. *Anthropologica, 40*(1), 49–70.

West, P. (2012). *From modern production to imagined primitive: The social world of coffee from Papua New Guinea*. Durham, NC: Duke University Press.

Chapter 12

Toward Economies That Won't Leave: Utilizing a Community Food Systems Model to Develop Multisector Sustainable Economies in Rural Southeastern North Carolina

Leslie H. Hossfeld, E. Brooke Kelly, Amanda Smith, and Julia F. Waity

Introduction

Food and place are inextricably tied together. The place where food is produced, who produces it, and the folkways and customs that surround the food we eat are central to all regions. North Carolina is no different: small family farms once dominated the North Carolina landscape, particularly in the southeast part of the state, where the rich, fertile soil made farming central to community life. Southeastern North Carolina has a long history and connection with agriculture, which served as the primary economy in many counties for decades. Food production has shaped the region and its rural communities and is deeply embedded as part of its history, economy, and culture. While agriculture still exists in this area, its form has changed. Replacing the small family farms that were intrinsically linked to the community and land are large agribusiness industries that have few ties to the region beyond economic incentive.

Poverty is also a deeply embedded and defining characteristic of this corner of the state. Once referred to as the "vale of humility between two mountains of conceit," (quote attributed to Zebulon Vance, in reference to the state's location between Virginia and South Carolina), North Carolina transformed itself from its humble origins to a progressive state embracing the new millennium. From the boom of the Research Triangle to the financial banking hub of Charlotte, North Carolina stood out on many indicators of progress, prosperity, and leadership. Yet the problems that have plagued the state for centuries—persistent poverty, low incomes, and racial inequalities—endure, and their residue is the very issue facing southeastern North Carolinians. The rural, small-town South captures much of what southeastern North Carolina has traditionally been, and is, today.

This chapter examines a place-based strategy focused on local food systems aimed at alleviating the high rates of poverty and food insecurity in our region. By strengthening the economy around a local food systems movement, these programs can bolster economies that *won't leave*. We begin by detailing some of the key terms and concepts that we will use throughout the chapter, including spatial inequality and food insecurity. Next, the region of southeastern North Carolina is described in detail, with emphasis on the importance of *place*. Then we turn to the demise of the small family farm with the associated rise in large agribusiness. We then use interviews with farmers in the region to illustrate some of the challenges they face and the successes of local food systems. The rest of the chapter is devoted to an alternative food movement, Feast Down East, which was created in response to the issues that southeastern North Carolina has faced, especially loss of jobs, high rates of poverty and food insecurity, and the decline of small family farms. We conclude with how Feast Down East has reclaimed *place* in southeastern North Carolina.

Spatial Inequality and Food Insecurity

Geographic place is important when studying inequality (Gans, 2002; Gieryn, 2000; Lobao & Saenz, 2002; Milbourne, 2010; Slack, 2010; Tickamyer, 2000). Instead of our traditional conceptions of inequality as *who gets what and why*, spatial inequality changes the concern to *who gets what and where* (Lobao, Hooks, & Tickamyer, 2007). Spatial inequality needs to be considered in addition to inequalities with which we may be more familiar, including race, class, gender, age, and sexuality. There are three ways of looking at the impact of place: compositional factors consider the characteristics of those living in a specific area; contextual factors consider the opportunity structures in the specific area; and collective factors focus on sociocultural explanations (Macintyre, Ellaway, & Cummins, 2002).

Rural areas differ from suburban and urban areas and, as a result, may offer different explanations for phenomena such as poverty and food insecurity. Rural residents are not only more likely to be in poverty, but also rural poverty is more persistent over time (Fisher, 2007; Hirschl & Brown, 1995; Tickamyer & Duncan, 1990). Rural and urban areas have both been devastated by structural factors, such as job loss, which contribute to poverty. As Sherman (2009) found in her study of a rural poor community that lost many logging jobs, welfare programs that encourage work are ineffective if there are no jobs available. Wilson (1996), researching in urban areas, found it was hard for residents to get ahead because of the opportunity structure in poor ghetto areas.

Place-based differences are evident when one considers food insecurity. Food insecurity is defined as a lack of access to enough food at all times for all members of the household to be healthy and active (Coleman-Jensen, Nord, & Singh, 2013). In 2012, 14.7 percent of Americans were food insecure. Of those, 5.6 percent reported very low food security, which indicates reduced meals and disrupted eating patterns, including hunger. Those living in suburban areas (outside principal cities in metropolitan areas) had a food insecurity rate of 12.7 percent, rural areas (outside metropolitan areas) had rates of 15.5 percent, and principal cities had rates of 16.9 percent (Coleman-Jensen et al., 2013). Previous research has shown that those with low incomes or low levels of education, single mothers, those who are black, Hispanic, or Native American, those who have three or more children, and/or those who live in a central city area are more likely to be food insecure (Nord & Parker, 2010). Food insecurity is associated with negative health outcomes. Adults who are food insecure have increased rates of morbidity, mortality, obesity, heart disease, diabetes, high blood pressure, poor self-reported health status, depression, and other mental health issues (Bhattacharya, Currie, & Haider, 2004; Siefert, Heflin, Corcoran, & Williams, 2004; Vozoris & Tarasuk, 2003).

Food insecurity is of particular importance to North Carolina, whose food-insecurity rate is 17.0 percent, the fifth highest in the nation, well above the US average of 14.5 percent (Coleman-Jensen et al., 2013). North Carolina's food-insecurity rate increased greatly as a result of the Great Recession (4.7 percent), and it still has not subsided to prerecession levels (see figures 12.1 and 12.2). Three southeastern North Carolina counties have food-insecurity rates above the North Carolina average: Bladen, Columbus, and Robeson Counties (Feeding America, 2011).

Food insecurity is often associated with living in a food desert. A food desert is defined by a lack of access to grocery stores and easier access to fast food and convenience stores, which generally have fewer healthy and nutritious options (USDA, 2009). The 2008 Farm Bill included language that defined a food desert as an "area in the United States with limited access to affordable and nutritious food, particularly such an area composed of predominantly lower income neighborhoods and communities" (Title VI, Sec. 7527, in USDA, 2009, p. 1). Food deserts can lead to higher rates of obesity and other diet-related diseases (Schafft, Jensen, & Hinrichs, 2009). While food deserts may not directly cause food insecurity, they are good indicators of areas where food insecurity is more likely (Morton, Bitto, Oakland, & Sand, 2005). Many factors increase the likelihood of an area being a food desert, including high numbers of residents with low income or poor education, low levels of vehicle ownership, high rates of poverty, high rates of unemployment, high rates of minority population, and high rates of vacant housing (Dutko, Ver Ploeg, & Farrigan, 2012). These factors affect rural and urban areas differently, with poverty being a stronger predictor of food deserts in urban areas, and vacant housing being significant in rural and less densely populated urban areas (Dutko et al., 2012). The elderly, the disabled, and those with low incomes are more likely to be affected by food-desert conditions.

Many areas in southeastern North Carolina are considered food deserts. All but one county (Pender) have at least one census tract that is considered a food desert. Robeson has the most food deserts at seven, with New Hanover close behind at six. Areas considered food deserts are often associated with higher rates of food insecurity. By focusing on the local food system, we can work to alleviate the food insecurity—and the often accompanying high poverty rates— associated with living in a food desert.

The Great Recession has had a profound effect on a large proportion of Americans, not just the poor or working class, but also the middle class. Additionally, this recession has exacerbated persistent poverty in southeastern North Carolina. In times of economic crisis, access to basic needs is sometimes lost. This is often referred to as material deprivation, which includes lack of access to shelter, food, clothing, and basic medical care. Households experience material hardship when they do not consume the minimum goods and services that are essential to survival (Beverly, 2001; Ouellette, Burstein, Long, & Beecroft, 2004). The recession increased need across the board, not just for those who were already poor. Focusing on material deprivation is critical because this measure captures those who may not be captured using traditional income measures of poverty. One important aspect of material deprivation is food insecurity.

Place and Southeastern North Carolina

Place, and more specifically region, is especially important when considering the South (Falk, Talley, & Rankin, 1993; Lyson & Falk, 1993). The Black Belt, located in the rural south, has been considered a "forgotten place" because of the lack of economic development that has occurred there (Falk et al., 1993). Our focus region of southeastern North Carolina, which is part of the Black Belt, contains a large swath of rural poverty that at times seems forgotten. The sense of place and community found in the "forgotten place" of southeastern North Carolina creates a unique context where a local food systems movement can help to address the distinct challenges faced by this area as a result of job loss and high rates of poverty and food insecurity. The conditions, losses, and structure in agriculture in southeastern North Carolina have created a major social problem in the region and across the state. With rising poverty and the massive loss of manufacturing and agricultural jobs, the entire rural economy of southeastern North Carolina was in need of reconstruction.

Southeastern North Carolina is a large rural region with one major urban center, Wilmington, and is part of the I-74 corridor between I-95 and the coast. The USDA considers Bladen, Columbus, and Robeson to be counties of persistent poverty, with poverty rates well over 20 percent since 1970. According to annual census data, Robeson County ranks among the top ten poorest counties in the nation (among counties with populations between 65,000 and 250,000); indeed over 33 percent of the county lives in poverty. The North Carolina Department of Commerce categorizes counties based on economic need, ranging from Tier 1, with the greatest economic need, to Tier 3, with the least need. There are five Tier 1 counties in southeastern North Carolina: Bladen, Columbus, Scotland, Hoke, and Robeson.

Southeastern North Carolina is the most ethnically diverse, multicounty region in rural America with the largest Native American populations East of the Mississippi River (Lumbee, Coharie, and Waccamaw-Siouan), and large numbers of African American, Hispanic, and European American populations. Southeastern North Carolina is also one of the three major regions of persistent poverty in North Carolina.

One contributor to the region's persistent poverty is farm loss. The seventh congressional district, serving all of southeastern North Carolina, lost 54,866 acres of farmland between 2002 and 2007 (USDA Department of Agriculture, 2007 US Census of Agriculture).

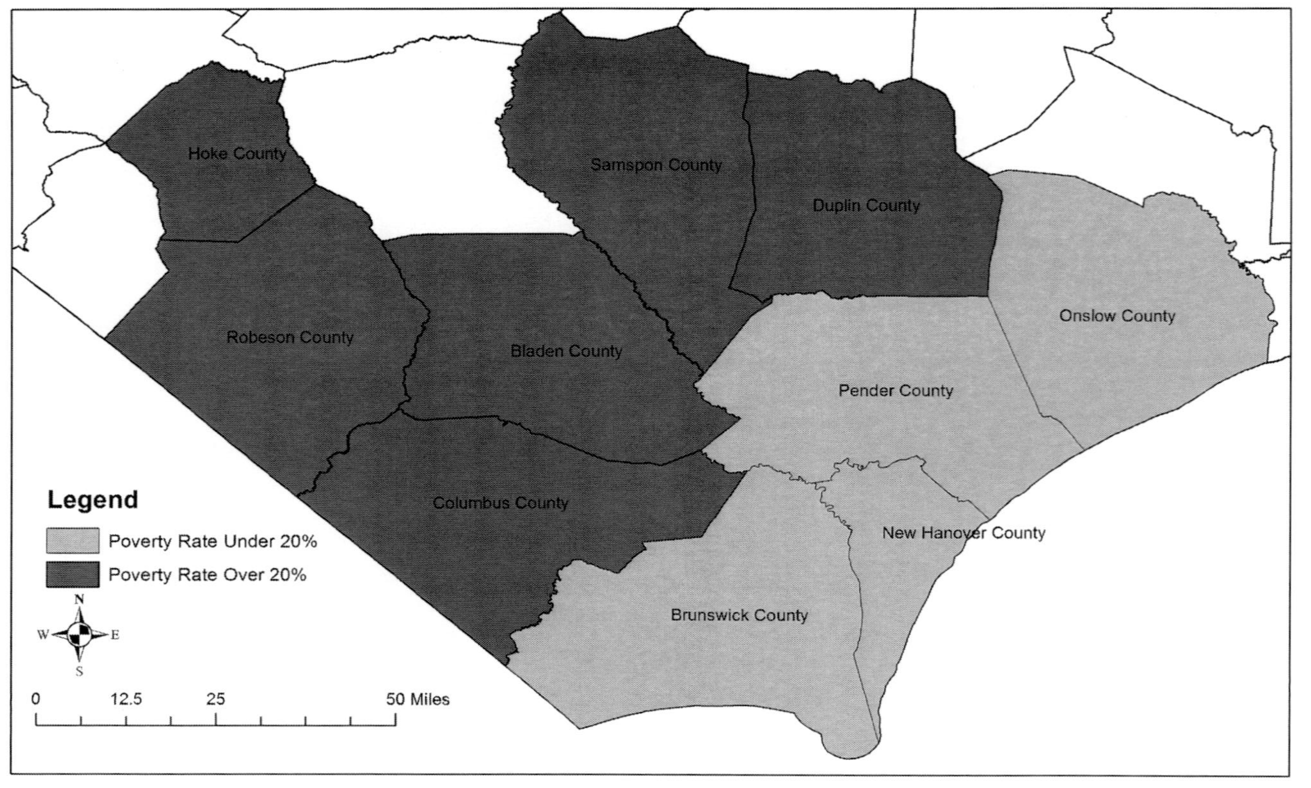

Figure 12.1 Southeastern North Carolina Poverty Rates

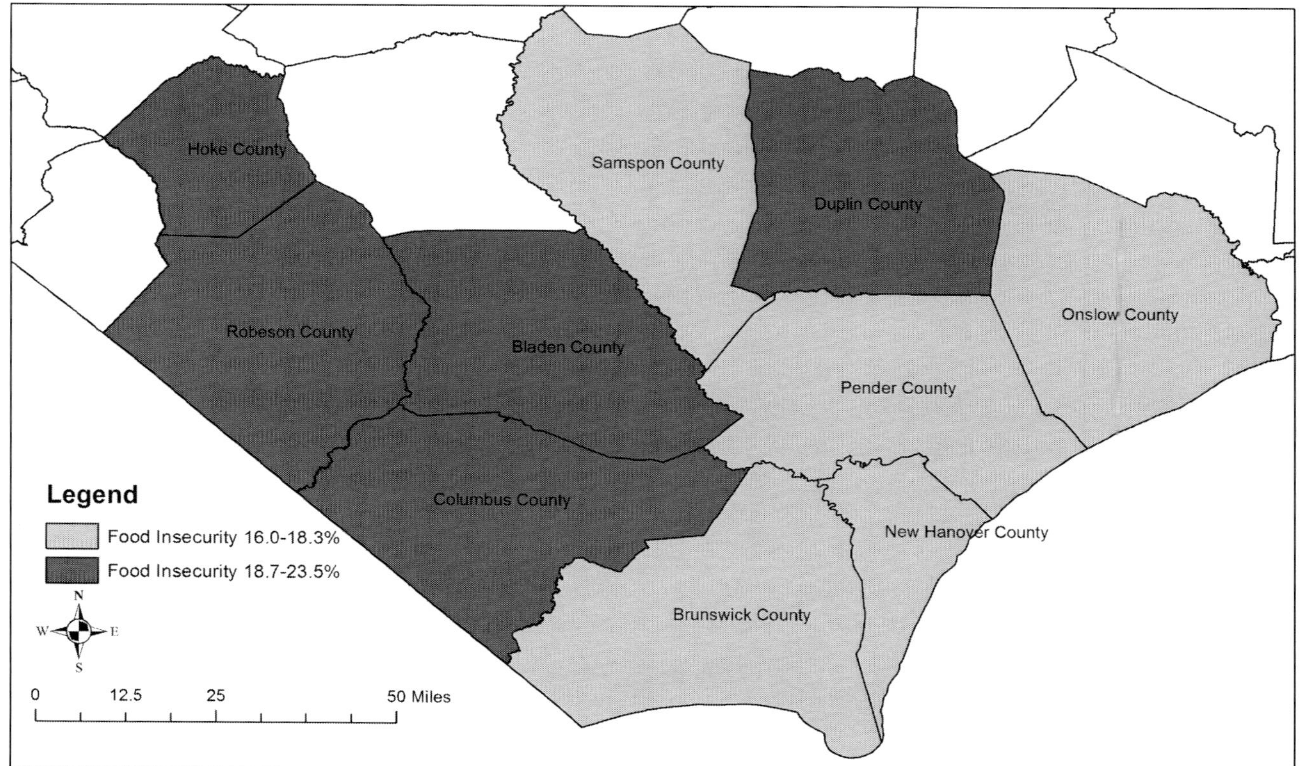

Figure 12.2 Southeastern North Carolina Food Insecurity Rates

North Carolina has lost more farms than any other state in the nation. In spite of this loss, the seventh congressional district continues to rank first in agricultural sales in North Carolina, with the total value of agricultural products sold being at $2,520,862.00 (2007 Census of Agriculture). This district ranks as the twenty-sixth most productive congressional district in the nation. Despite that fact, 60 percent of the farms (2,905 of the 4,809) in the seventh district had less than $20,000 in farm sales in 2007, indicating that these were smaller family farms. These limited-resource farmers are aging and have few health and retirement benefits, which may cause them to give up their farms and contribute to farm loss. At the same time, low-income consumers in this region are also aging, growing in number, and clustered in twenty-nine food deserts in the eleven-county region. A new model was needed to encourage small family farms to remain in business and also to create a path for the local produce to travel from those farms to low-income consumers.

Demise of Family Farms

Since 1935, the number of farms in the United States has decreased from 7 million to 2.2 million in 2012 (USDA, 2012). Today farmers make up only 2 percent of the population. Much of the decline occurred as a result of the emergence of urbanization, global food systems, large-scale industrial agribusiness, and changes in agricultural policies. Farmers today are not only competing with other farmers in their nation, but also with a market that sources from producers worldwide. Scholars argue that the agricultural sector has experienced a "race to the bottom," following other globalized sectors (e.g., manufacturing and textiles) through a new international division of labor (McMichael, 2000). Since the 1970s, agricultural production and distribution has progressively transformed into a global enterprise characterized by industrialization and openmarket policies. Food systems are now global phenomena and agriculture a "world industry" (McMichael, 2008, p. 106). Today, one in three acres of crops are planted for export, and 31 percent of US farm income comes from exports (American Farm Bureau Federation, 2015). The global market is dominated by free-trade philosophies that focus on bottom-line profits and the assertion that farmers will survive competition if they are efficient and seek comparative advantage (McMichael, 2005). However, today's food systems are no longer based on competition between scores of farmers but, rather, are controlled by a few dominant agribusiness entities (McMichael,

2005; Pingali, 2006; Welsh, 2009). Indeed, farms with gross sales exceeding $250,000 equal 12 percent of all farms, yet this percentage controls 84 percent of food production in the United States (Hoppe & Banker, 2010).

The rise of large-scale, industrial agribusinesses has resulted in a system that is dominated by just a handful of powerful firms controlling food systems, setting prices and influencing policy, and removing small producers from the decision-making process altogether (Hendrickson & Heffernan, 2007). Subsequently, small farmers in these markets find it increasingly difficult to compete. What McMichael (2005) terms the "corporate food regime" is a global food system that is controlled, not by citizens, nor the vast majority of farmers, but a small number of corporate firms and industrial farms. This places small farmers and rural communities in subordinate positions with limited control over the policies and processes that directly affect their lives. At the heart of this transition lie the countless rural communities that depend on agriculture and food production as their primary economies.

Importance of Small-Farm Agriculture for Rural Communities

> *Having fifteen farms here vs. having one big farm—it hurts the economy because it takes away jobs from the area. If you had fifteen farms they could employ more people because the large farms do everything with machinery. So it puts some people out of jobs.* [Feast Down East Farmer]

Previous studies have outlined the damage done to rural communities when agriculture leaves (Durrenberger & Thu, 1996; Heffernan, 2000; Swenson, 2009). Moreover, research suggests that it is not just any form of agricultural production that stimulates rural communities, but the presence of small farms that counts (Lobao & Meyer, 2001; Lyson & Guptill, 2004; Welsh, 2009). Studies done in the United States have shown that local economies in areas where corporate farms replaced smaller family farms typically suffer, as the monies made at a corporate farm are often funneled back to a corporate headquarters and away from local communities (Durrenberger & Thu, 1996; Heffernan, 2000; Swenson, 2009). Conversely, areas with a significant number of small family farms showed a greater percentage of income being put back into local economies through the purchase of farm inputs and household goods, thus promoting the overall economic well-being of these communities (Heffernan, 2000). A similar argument, the Goldschmidt hypothesis, asserts that the increasing presence of large-scale industrial farms polarizes rural communities. The hypothesis suggests that as large farms replace small farms, farmers lose control over the means of production and instead begin selling their labor, resulting in a shrinking middle class of farmers (Welsh, 2009).

Alternative Models

> *Right now local is being pushed and it needs to be because we don't need to truck stuff from overseas and California because we can grow it all here if we got it all lined up in this region. We have enough land.* [Feast Down East Farmer]

Whereas globalization and industrialization place small farmers and rural communities at a disadvantage, alternative models, such as local food systems, are thought to provide fair access to markets and to stimulate local economies. These systems work toward a recovery of the local, seeking to reconnect people with the origins of food and to bring citizens closer together. Local systems focus on strengthening local economies, poverty alleviation, and community-food security. They operate on the premise of short supply chains based not only on the shortening of physical distance, but also on the networks of actors that create "socially embedded" systems of exchange (Hinrichs, 2000; 2003).

Local food systems often go beyond the economic advantages for producers, positively impacting the economies in which they are situated. As more local food is purchased, more money is in the hands of the farmer, who, in turn, spends on inputs and other supplies from surrounding businesses (Durrenberger & Thu, 1996; Heffernan, 2000; Rosset, 2000; Swenson, 2009). Earlier research has demonstrated the significant economic impacts that local systems have on communities through job creation and economic stimulus, using input/output and multiplier models (Swenson, 2009). Small farms also depend on local systems, as a large majority of their sales are through local markets (Low & Vogel, 2011).

Local systems have become increasingly popular over the last two decades. To illustrate, the number of farmers markets in the United States has gone from 1,755 in 1994 to 7,864 in 2012 (USDA, 2012). Likewise, local food sales have risen from $551 million in 1997 to $4.8 billion in 2008 (Low & Vogel, 2011; Martinez et al., 2010). A significant increase in Community Supported Agriculture (CSA) is also noted. CSAs are partnerships between consumers and farmers where consumers cover anticipated costs of food production in exchange for a share of the farm's produce (DeMuth, 1993). According to the USDA (2010), there were only 400 CSAs in the United States in 1984. By 2010, the USDA estimated the figure to be 1,400. However, since tracking of CSAs is limited, the estimated number is expected to be much greater. Although local food sales and systems continue to gain in popularity and sales, they comprise only about 2 percent of all agricultural product sales.

Farmers working in local systems and consumers purchasing from local systems are working to reverse this trend. A reclaiming of place is underway, where economies are being reconstructed, cultural practices reborn, and communities are once again coming together around food and agriculture. While the popularity of the local movement in urban areas is relatively new, local food and farming in rural communities is tradition. The movement is only a resurgence of a previous era and a push to revitalize the region's natural economy.

Toward Economies That Won't Leave: The Development of Feast Down East

Federal trade policies, the outsourcing of manufacturing, and the demise of family farms have fundamentally changed rural economies across the United States. Although policy makers acknowledged that certain communities and regions would be disproportionately impacted by these shifts, rural communities across the United States have been left largely on their own to reconstruct their economies and societies in a systematic way. No state or federal programs have been specifically developed to identify the most impacted counties and to develop partnerships with resources to reconstruct rural economies and societies. Southeastern North Carolina is no exception to this, having been hard hit by manufacturing textile job loss and the erosion of small family farms. The diverse peoples and communities of southeastern North Carolina face challenges but have multiple opportunities to improve the economic and social conditions and quality of life in their communities. The decision to form Feast Down East was based on the realization that agriculture and manufacturing are the two most challenged economic sectors in our region. There are opportunities and evidence-informed solutions to the major challenges in both sectors. Local food production is recognized as the most effective solution to developing a sustainable food system in rural and urban communities across North Carolina, the United States, and the world.

The Southeastern North Carolina Food Systems Program (SENCFS)—also known as Feast Down East—began in 2006 as an economic and community-development initiative in response to the growing poverty rate and the massive job loss in the region's agricultural and manufacturing sectors. Through grassroots organizing and an intentional focus on poverty alleviation and community and economic development, Feast Down East has advanced into a partnership of public and private institutions and agencies in eleven rural and urban counties. Its aims are: to maximize market opportunities; to keep a greater percentage of the food dollar within southeastern North Carolina; and to increase local and regional wealth through the multiplier effect of expanded markets, sales, and profits.

The governance of Feast Down East is democratic, farmer driven, and supported by public and private service providers and businesses. The Feast Down East Advisory Board is diverse in terms of race, geography, and sector and includes farmers, institutional buyers, educators, and policy makers. The board is representative of the demographic composition of southeastern North Carolina. Cooperative extension agents in each county are directly engaged in local and regional planning, service provision, and governance.

The mission of Feast Down East is to connect institutions, agencies, farmers, businesses, and consumers; to support, coordinate, expand, and sustain the production and consumption of healthy local foods, particularly by, and among, limited-resource farmers and limited-resource consumers; to create an economically viable, regional food-system and public/private partnerships that benefit farmers, businesses, food services, and consumers; and to ensure access to healthy, affordable food by all consumers in southeastern North Carolina. Feast Down East is principally committed to increasing the capacity of limited-resource farmers in becoming resourceful farmers and in supporting limited-resource communities in advancing their own food security.

Feast Down East completed three years of research and local food assessments that identified seven major elements and needs in a regional food system in southeastern North Carolina. These are: (1) profitable private and public markets for local food sales; (2) comprehensive support for, and engagement of, limited resource farmers and measurable outcomes to becoming resourceful farmers; (3) the processing and distribution of local foods for year-round sales and consumption of healthy foods; (4) a highly diverse and strong public/

private partnership; (5) food security and engagement of low- and moderate-income consumers in the twenty-nine food deserts in the region; (6) the establishment of Food, Farm, and Family Councils (adapting a Food Policy Council model) that engage all stakeholders in the coordination of local food production, processing, distribution, sales, and consumption; and (7) significant public and private financial and nonfinancial support.

The goals of poverty reduction, engagement, and empowerment of consumers and limited-resource farmers (defined by the USDA as socially disadvantaged farmers; in our region this is primarily African American and women farmers) are the foundation and the beneficiaries of the system's development and programs. Feast Down East has created a comprehensive regional food system based on key partnerships in its eight rural and three urban counties. Major partners and their roles are:

- North Carolina Cooperative Extension Service, which provides farm support services including Good Agricultural Practices training as well as nutrition programs for low-income consumers
- community colleges—small-business training providers
- child-nutrition directors in public schools and universities—food purchasers and preparers
- the Wilmington Housing Authority, which creates programs that provide training and nutrition classes for low-income consumers in eight food deserts
- Aramark, the local food-service provider for the University of North Carolina–Wilmington, and food purchaser and buy-local educator
- Feast Down East Processing and Distribution Program, with forty limited resource farmers—producers and distributors of local, healthy food
- Center for Community Action, cofounder and anchor partner with nationally recognized experience in rural development
- farmers markets and direct farm-product outlets with Feast Down East as agent for eight markets
- Feast Down East Food Corps service member for Brunswick and New Hanover counties working in schools to expand nutrition awareness and community gardens
- and Feast Down East VISTA service members, who address food insecurity in the region.

Low-income consumers and farmers have been actively engaged in Feast Down East since its inception. Individuals and organizations involved in Feast Down East attend monthly planning and implementation meetings in the Wilmington area. Food, Farm, and Family Councils have been established in the rural counties. Feast Down East's limited-resource farmers, predominately African American and women farmers, lead and benefit from the work of the program's local food processing and distribution center.

Our greatest concern and critique of the national local food movement is that it has mainly become an experience for middle-class consumers. Because of this, Feast Down East created its Food Sovereignty Program whereby limited-resource consumers increase their access to fresh and affordable local food, gain knowledge and skills in developing and managing food via buying clubs, and take nutrition classes in direct cooperation with producers. Through this program, limited-resource farmers increase their revenue and acquire additional skills and supports needed to become resourceful farmers. Limited-resource consumers also learn leadership skills in coordinating their buying clubs, nutrition classes, and direct engagement with farmers and their farms. The project transforms the entire relationship of low-income consumers with their own health, the food that they consume, and the farmers and farms that source their food. The Feast Down East Food Sovereignty Program builds a circular system of mutual support and sustainability that improves the well-being of their livelihoods and community.

Feast Down East supports limited-resource farmers in meeting living-income standards, providing sustainable livelihoods, and moving out of poverty. We have learned that if a regional food system hopes to alleviate poverty, a specific focus on supporting limited-resource farmers in becoming resourceful farmers is paramount. Women and minority-owned farms are small and more diverse; and having been marginalized by the big-agribusiness model, they are more likely to participate in the local food movement. They are especially in need of programs such as Feast Down East. We have learned that these three elements are critical: an organized and effective system of institutional buying that provides profitable markets for the purchase of local foods; an organized and effective infrastructure to support local and regional food production, including public and private service providers, educational institutions, and consumer groups, both inside and outside the agriculture sector; and an organized and effective system of nonfinancial/financial support for farmers, to enter or transition to local food production. Furthermore, we believe limited-resource food desert consumers should benefit from the local-food movement through targeted projects such as the Feast Down East Food Sovereignty Program, which ensures access.

Farmers' Perspectives: Challenges and Successes in the Local Food System

Since agriculture is so pivotal to the economy of this region, we focus on farmers' experiences as a way to illustrate some of the challenges and resiliencies of operating within local food systems. Interviews with limited-resource famers at various stages of local food systems participation were conducted. Interviews included questions about the farmers' practices and challenges, and they ranged from approximately twenty minutes to an hour.

To highlight the challenges that farmers face in two high-poverty counties where a local food movement is in its infancy, twenty-seven interviews with farmers in Robeson and Columbus Counties (referred to as Robeson County for simplicity) are addressed. Drawing on interviews with thirteen Feast Down East participating farmers in New Hanover, Pender, Duplin, and Sampson—where a socially embedded local food system has flourished—we then illustrate the significance of local food systems and, more specifically, Feast Down East, in addressing such challenges.

Challenges of Robeson County Farmers

Robeson County has a rich history in agriculture, though farming has declined and shifted to larger, less diversified approaches, as described

Table 12.1 Demographics of farmers

Age	Robeson County	Feast Down East
25 or younger	2	0
26–40	5	1
41–59	9	5
60 or older	11	7
Race/Ethnicity		
American Indian	13	0
African American	1	5
White	11	8
Gender		
Women	3	5
Men	24	8
Years farming		
Average	32	15
Range	61	33
Total in sample	27	13

by one Robeson County farmer: "I used to be in produce really heavily: squash, tomatoes, cucumbers, watermelons, sweet corn, okra, and butter beans. I kind of switched to row cropping. It was easier, less help, main thing was the labor expenses... We pick everything with a machine rather than manual labor." Farmers who are returning to, or have struggled to remain small-scale with local sales, face particular challenges in doing so. Indeed, farmers often addressed labor as a challenge (eight out of twenty-seven), but they most frequently addressed marketing (nineteen out of twenty-seven). Feast Down East addresses many of the challenges Robeson County farmers raised by fostering a local food system that supports small-farm agriculture through an alternative model.

Challenges of Marketing in High-Poverty Context

A tough economy in an area with a high percentage of individuals below the poverty line can exacerbate the challenge of marketing to local consumers, as described by Robeson County farmers: "We're in a poor county, and it's a little bit harder to move [our crops] because of the fact that we gotta have that premium price." Another farmer further elaborates: "We live in... a very frugal county... What I mean by that is people don't want to spend retail prices at a farmers market or a little roadside stand. They'll go to the grocery store and pay ninety-nine cents a pound for collard greens, when they can buy a five-pound collard out here for two dollars." He further explains later in the interview, "Our economy has really made things a lot more difficult for the public. The people are looking for deals, and they're looking to buy things at one place." With consumers compelled by economic forces and poverty to find less expensive products at superstores, four farmers addressed the need to educate consumers about the benefits of purchasing locally grown food.

Marketing

One of the most common barriers for farmers across the region is breaking into new markets. Farmers noted several difficulties. Product selection and price setting were frequently cited as initial barriers when trying to establish new markets. That is, many farmers were unsure of what to grow, the appropriate price to set for the product, and the most suitable market outlet. to avoid a surplus of produce, farmers must establish a market for their crops before they are planted. This is problematic when market opportunities are not readily apparent or available.

In open-ended questions about challenges that they face, Robeson County farmers described the need to "find buyers," "get a market base," and/or "get my name out to buyers." One Robeson County farmer described the difficulty of switching gears from their everyday tasks to marketing the crops they produce: "We're farmers and we don't completely understand marketing all the time... When you get out of a piece of equipment and, uh, leave from the farming aspect, as far as producing it and raising it... It's just a completely different ball game, and we're not used to that." Another Robeson County farmer further elaborated on the challenges of anticipating consumer demands: "Trying to figure out what people's seasonal habits are [is a challenge]. I can grow anything but don't know what people want or need."

Problems with Timing of Crops, Supply, and Demand

Eight Robeson County farmers discussed challenges with the timing of crops, such as trying to plan and sell at the right time to optimize freshness and profit. In addition to discussing problems with what crops to grow, when, they also addressed the challenge of how much to grow. Farmers described problems with not having enough yields to meet demand, as well as occasions in which they had more than they could sell before the product spoiled: "[The challenge is] trying to sell it at the most profitable period of time and get the better price for it... If your timing is correct with tomatoes, cantaloupes, watermelons, okra, and things of that nature you get a good/better price for it... Those things are delicious when they are grown and produced locally and [when] you sell to a customer they can really get the real taste of the product... versus buying something from Florida... and the taste... the real quality of the product is not there." Another farmer elaborated on the challenge (addressed by several farmers) of scrambling to put excess yields to use before they could spoil. He responded to a question about the challenges he faces as a farmer with the following: "Uh, choosing the correct product, making the decision about the quantity to grow and growing excess. I'd rather have less amount than to have excess. So sometimes I have excess, [that] is when I have to go get on the phone and maybe peddle... But that's the biggest challenge is deciding how much to grow." In such circumstances, some farmers discussed trying to "give it to somebody if I can't sell it before it gets unusable." Not only do such circumstances present the possibility of lost revenue for farmers, but excess yields also hold potential for gleaning in a high-poverty context with high food-insecurity rates.

Increased Access to Markets through Local Food Systems Participation

Farmers with access to a network of other farmers participating in local food systems, specifically Feast Down East, reported that these connections significantly aided in the issues highlighted from the Robeson County interviews. Farmers were able to plan more crops according to the markets established through Feast Down East. It is noted that many of the smallest, most limited resource farmers used Feast Down East as their primary market outlet. Many lacked other means of marketing their products and/or did not have the time and resources to participate in direct sales activities. This is particularly important, as most limited-resource farmers often struggle to break into sales markets. Many farmers noted that before becoming involved with Feast Down East, they experienced difficulty locating sales outlets and often found themselves with a surplus of harvested product. As one Feast Down East farmer notes: "When we first started we gave a lot of stuff away because we didn't have anybody to sell it to... We didn't have a market for it and were just out there doing our own thing. Our first year we had so much corn, and when we got it in we didn't have anybody to sell it to."

Once established and working within the local system, farmers typically expressed a shift in focus to solidifying a consumer base. Farmers recognize that their consumer base is not the majority of the population. They often cited no desire to compete with larger distributors, nor to be situated in dominant markets. Instead, participating farmers focused on building a direct connection with buyers interested in farm fresh local. As noted by one Feast Down East farmer: "I'm not going to be able to grow five million pounds of turnips like a big producer, so I don't even try. I grow specialty crops. I grow what the local consumer wants. A special crop they are willing to pay more for—farm fresh."

Need for Consumer Education

Several Robeson County area farmers connected the tendency of local consumers to purchase less expensive nonlocal food with a need for education about the benefits of buying locally grown produce and meat. One farmer explained: "I think we're probably looking at the education for the consumers because there is a lot of misunderstanding about pricing because our chickens are more expensive than what you can get in the stores. But our chickens take... about twenty-four weeks or more to grow to full size. And the ones that you buy in the store take six to eight weeks to grow to full size before they are processed. So, a lot of people don't understand why our chickens are expensive. I can't make it any less expensive. I haven't raised my price on the chickens in three years."

Another Robeson County-area farmer further elaborated that the biggest challenge in selling his produce was "showing and educating others [about] the comparison [between] buying produce in a grocery store versus a farmers market. A farmer selling produce is very valuable and more profitable because the produce is fresh, not waxed and polished, more nutritious, and not refrigerated."

Several Robeson County-area farmers addressed the need for more farmers to grow produce locally. In discussing the need to educate local consumers, farmers proposed solutions such as increasing involvement of extension and other programs to promote the purchasing of locally grown food, educating young people who may not know about the source of their own food, and cooking demonstrations and classes to showcase the benefits of locally grown, fresh food.

Increasing Consumer Awareness and Demand of Local Products

Marketing efforts and buy-local initiatives were frequently cited as a primary benefit of working with local food system organizations. Farmers working with Feast Down East commonly referenced how these campaigns were raising consumer awareness in the region, thus impacting consumer demand for local food. As one farmer noted, "They [Feast Down East] have created a demand for it by showing it is a good thing to have and to use and therefore they are expanding the fresh vegetable market around here." Another farmer asserted: "There is plenty of room for farmers like me here. It's just getting more folks on board with buying local and being conscious of that. And that is where working within the local food systems is definitely beneficial—the advocacy. Having the big Feast Down East billboard on Oleander (Drive)—It's like 'buy local!'"

Some farmers also indicated an increase in consumer awareness specifically for their farm. That is, participating farmers have experienced an increase in farm name/brand recognition in the wider community. This is not surprising, as local systems are often based on building connections between producers and consumers. Many farmers noted the organization's marketing efforts as being pivotal in reaching wider consumer bases and increasing farm revenue. Several farmers recognized the role of Feast Down East buy-local initiatives in increasing sales in other markets, most notably through direct-to-consumer outlets (e.g., farmers markets and farm stands), as more consumers are becoming familiar with the area's local farmers.

Social Embeddedness of Economic Transactions within Local Food Systems

In addition to the aforementioned findings, the interviews with farmers participating in local food systems revealed that economic transactions were embedded within social relationships with buyers: "The local food system is your community of peers as a small farmer, be it your customers, other farmers, people advocating local food, restaurants advocating local food... It is like a big spiderweb of people and, you know, they just exchange ideas, and money and products. It's its own niche economy." For those participating in local food systems, many economic activities took place within relationships typically forged through the common ideological goals of the local food movement, the commitment to local communities, and the shared history of the region. Local movements are often built around the relationships developed between farmers, community members, and local businesses. Activity in alternative systems often goes beyond rational economic transactions and focus on principles such as trust, loyalty, and commitment. As one Feast Down East farmer noted, "We aren't going to pull one over on our customers to make a dime. Keeping the relationship is more important to us than the money."

Farmers who are able to build relationships with buyers, based on common beliefs about the local movement, or who can work out of existing relations, tend to fare better in local food systems. For instance, these direct partnerships often provide farmers with stable sales outlets that aid in crop planning and help provide market security. As a Feast Down East farmer shared: "Mark [pseudonym] told me he would buy from me before he even opened his restaurant, and he has stuck by his word and buys from me every week... I have developed a really good relationship with that restaurant in particular. I grow stuff specifically for them and if I have extra I sell it elsewhere."

These relationships not only increase the chances of repeat customers, but they also strengthen and shape networks in local food systems that are accessed for future social and economic capital. Although many farmers stated that they were using their existing social capital and networks to help connect with other actors in local food systems, they also noted the importance of local food systems organizations in establishing new connections and bridging social capital. As noted by one Feast Down East farmer: "It was hard to get into the couple of restaurants we are in, and actually Feast Down East helped with that. Once we got in, eventually when we have more product, we are going to go to that chef and say give me a name. I have leftovers, so I need another name. And that is how you build up your connections. Once you establish with one, they can get you in with others."

The embeddedness of markets within the local system, and particularly within the region where Feast Down East is located, is the cornerstone of their success. They work not only to bridge the divide between producers and consumers, but to create new networks of actors invested in the local community. In turn, it is the nature of the communities that fosters this embeddedness, as most citizens, farmer and consumer alike, have a vested interest in seeing agriculture flourish in their region again.

Conclusions and Observations: Reclaiming Place

> *Well now there are two sets of farmers. The big farmers will continue on forever. The small farmers are the ones we are losing. You know this was an agrarian society from the 1800s to the 50s. Eighty to 90 percent of the people lived on the farm. General motors and all them people cranked out all the jobs up north and everybody left. And the small farms, people left them. But this is a new day. People are coming back. They are leaving the city and reconnecting. The world is changing so fast that a lot of people want to get off and come back and relax for a minute.* [Feast Down East Farmer]

What is unique about this region and its participation in the local food movement is its ties with agriculture. Even in the urban hubs where the local products of southeastern North Carolina are often sold, one would be hard-pressed to find a resident with absolutely no connection to farming and food production. Citizens of the region have a shared history that is built upon this connection to agriculture. Feast Down East works to rejuvenate these existing structures and relationships in an effort to revitalize the economy of the region. Even more important, local systems work to reverse the devastating trends that have eroded generations of culture and history surrounding food and food production, which are intrinsically linked to region.

Local food systems initiatives, such as Feast Down East, can work to correct issues with market access and fair prices, but just as important is raising consumer awareness on the importance of buying local for their region. Informed citizens, coupled with dedicated farmers, create the socially embedded economic systems necessary for these models. Without programs such as Feast Down East that focus on systemic *change*, we could eventually have a local food system made up of large farms that primarily serve middle- and upper-income families with no impact on poverty reduction and prevention and the economic development and security of the family farm.

Recent initiatives from the USDA point to this growing concern. In 2010, United States Secretary of Agriculture, Tom Vilsack, created the Strikeforce for Rural Growth and Opportunity, an initiative that helps develop programs and partnerships to assist persistent poverty in communities and underserved populations in rural America. A recent rollout of this initiative in North Carolina appointed Feast Down East to provide guidance on this initiative throughout the state because of its organizational principles of poverty alleviation and economic development.

Feast Down East has developed into a model food systems development and systems change organization based on access, equity, and inclusion. The program assumes that authentic development of communities and peoples must be directly linked to poverty reduction, dismantling institutional discrimination, promoting social justice, and empowering both grassroots and professional peoples and communities. While every community faces unique challenges and must tailor food systems projects to fit the composition and needs of their region, those interested in doing this work can find lessons and examples from Community Food Systems projects throughout the country. Feast Down East's success can be attributed to the key characteristics that focus on an "emphasis on strengthening existing (or developing new) relationships between all components of the food system" (Cornell, 2013, p. 1). Triple bottom-line approaches that include economic, environmental, and social justice goals help ensure the sustainability of community over time.

What we know is that reducing poverty is very complex because successful efforts must include a comprehensive approach with multiple strategies on multiple levels: there is no one solution, and multiple programs are needed in combination with one another to create significant impact. A comprehensive systems approach is necessary because the causes of poverty are rooted in multiple levels, including the lives of individuals, families, communities, institutions, systems, and culture. Communities may implement multiple programs, but if the projects are not placed in a larger context with significant program and policy alignment, program and policy development will be isolated, having little overall impact on poverty reduction and economic recovery in any given community, county, state, or region.

Feast Down East helps limited-resource farmers and limited-resource consumers reclaim *place*, building the capacity and well-being of *both* as beneficiaries of local food, and in doing so, creating economies that *won't leave.*

References

American Farm Bureau Federation (2015). Fast facts about agriculture. *The voice of agriculture.* Retrieved from http://www.fb.org/index.php?fuseaction=newsroom.fastfacts

Beverly, S. G. (2001). Material hardship in the United States: Evidence from the Survey of Income and Program Participation. *Social Work Research, 25*(3), 143–151.

Bhattacharya, J., Currie, J., & Haider, S. (2004). Poverty, food insecurity, and nutritional outcomes in children and adults. *Journal of Health Economic, 23*(4), 839–862.

Coleman-Jensen, A., Nord, M., & Singh, A. (2013). *Household food security in the United States in 2012* (ERS Publication No. ERR-155). Washington, DC: US Department of Agriculture, Economic Research Service. Retrieved from http://www.ers.usda.gov/publications/err-economic-research-report/err155.aspx

Cornell University. (2013). A primer on community food systems: Linking food, nutrition and agriculture. Retrieved from http://www.resilience.org/resource-detail/1442546-a-primer-on-community-food-systems

DeMuth, S. (1993). An excerpt from community supported agriculture (CSA): An annotated bibliography and resource guide. USDA National Agricultural Library.

Durrenberger, E. P., & Thu, K. M. (1996). The expansion of large scale hog farming in Iowa: The applicability of Goldschmidt's findings fifty years later. *Human Organization, 55*(4), 409–415.

Dutko, P., Ver Ploeg, M., & Farrigan T. (2012). *Characteristics and influential factors of food deserts* (ERS Report No. ERR-140). Washington, DC: U.S. Department of Agriculture, Economic Research Service. Retrieved from http://www.ers.usda.gov/publications/err-economic-research-report/err140.aspx

Falk, W. W., Talley, C. R., & Rankin, B. H. (1993). Life in the forgotten South:The Black Belt. In T. A. Lyson & W. W. Falk (Eds.), *Forgotten Places: Uneven development and the loss of opportunity in rural America* (pp. 53–75). Lawrence: University Press of Kansas.

Feeding America (2011). Map the meal gap. Retrieved from http://feedingamerica.org/hunger-in-america/hunger-studies/map-the-meal-gap.aspx#

Fisher, M. (2007). Why is U.S. poverty higher in nonmetropolitan than in metropolitan areas? *Growth and Change, 38*(1), 56–76.

Gans, H. J. (2002). The sociology of space: A use-centered view. *City & Community, 1*(4), 329–339.

Gieryn, T. (2000). A space for place in sociology. *Annual Review of Sociology 26*, 463–496.

Heffernan, W. D. (2000). Concentration of ownership and control in agriculture. In F. Magdoff, J. B. Foster, & F. H. Buttel (Eds.), *Hungry for profit: The agribusiness threat to farmers, food, and the environment* (pp. 61–75). New York, NY: Monthly Review Press.

Hendrickson, M., & Heffernan, W. (2007). *Concentration of agricultural markets.* Department of Rural Sociology University of Missouri. Retrieved from http://www.foodcircles.missouri.edu/07contable.pdf

Hinrichs, C. C. (2000). Embeddedness and local food systems: Notes on two types of direct agricultural market. *Journal of Rural Studies, 16*(3), 295–303.

Hinrichs, C. C. (2003). The practice and politics of food system localization. *Journal of Rural Studies, 19*(1), 33–45.

Hirschl, T. A., & Brown, D. L. (1995). The determinants of rural and urban poverty. In E. N. Castle (Ed.), *The changing American countryside: Rural people and places* (pp. 229–246). Lawrence: University Press of Kansas.

Hoppe, R., & Banker, D. E. (2010). *Structure and finances of U.S. farms: Family farm report.* US Department of Agriculture. Washington, DC: Economic Research Service. Retrieved from http://www.ers.usda.gov/publications/eib-economic-information-bulletin/eib66.aspx

Lobao, L., & Meyer, K. (2001). The great agricultural transition: Crisis, change, and social consequences of twentieth century US farming. *Annual Review of Sociology, 27*, 103–124.

Lobao, L., & Saenz, R. (2002). Spatial inequality and diversity as an emerging research area. *Rural Sociology, 67*(4), 497–511.

Lobao, L. M., Hooks, G., & Tickamyer, A. R. (2007). Introduction: Advancing the sociology of spatial inequality. In L. M. Lobao, G. Hooks, & A. R. Tickamyer (Eds.), *The sociology of spatial inequality.* Albany: State University of New York Press.

Low, S., & Vogel, S. (2011). *Direct and intermediated marketing of local foods in the United States.* US Department of Agriculture. Economic Research Report 128. Washington, DC: Economic Research Services. Retrieved from http://www.ams.usda.gov/AMSv1.0/getfile?dDocName=STELPRDC5097250

Lyson, T. A., & Falk, W. W. (1993). Forgotten places: Poor rural regions in the United States. In T. A. Lyson & W. W. Falk (Eds.), *Forgotten places: Uneven development in rural America* (pp. 1–6). Lawrence: University Press of Kansas.

Lyson, T. A., & Guptill, A. (2004). Commodity agriculture, civic agriculture and the future of U.S. farming. *Rural Sociology, 69*(3), 370–385.

Macintyre, S., Ellaway, A., & Cummins, S. (2002). Place effects on health: How can we conceptualise, operationalise and measure them? *Social Science and Medicine, 55*(1), 125–139.

Martinez, S., Hand, M., Da Pra, M., Pollack, S., Ralston, K., Smith, T., . . . Newman, C. (2010). *Local food systems: Concepts, impacts, and issues.* US Department of Agriculture. Washington, DC: Economic Research Services. Retrieved from http://www.ers.usda.gov/media/122868/err97_1_.pdf

McMichael, P. (2000). Global food politics. In F. Magdoff, J. B. Foster, & F. H. Buttel (Eds.), *Hungry for profit: The agribusiness threat to farmers, food, and the environment* (pp. 125–144). New York, NY: Monthly Review Press.

McMichael, P. (2005). Global development and the corporate food regime. *Research in Rural Sociology and Development, 11*, 265–299.

McMichael, P. (2008). *Development and social change: A global perspective.* Thousand Oaks, CA: Pine Forge Press.

Milbourne, P. (2010). Putting poverty and welfare in place. *Policy and Politics, 38*(1), 153–169.

Morton, L. W., Bitto, E. A., Oakland, M. J., & Sand, M. (2005). Solving the problems of Iowa food deserts: Food insecurity and civic structure. *Rural Sociology, 70*(1), 94–112.

Nord, M., & Parker, L. (2010). How adequately are food needs of children in low-income households being met? *Children and Youth Services Review, 32*(9), 1175–1185.

Ouellette, T., Burstein, N., Long, D., & Beecroft, E. (2004). *Measures of material hardship: Final report.* U.S. Department of Health and Human Services. Retrieved from http://aspe.hhs.gov/hsp/material-hardship04/

Pingali, P. (2006). *Agricultural growth and economic development: A view through the globalization lens.* Presented at the 26th International Conference of Agricultural Economists. August 12–18, Gold Coast, Australia.

Rosset, P. M. (2000). The multiple functions and benefits of small farm agriculture in the context of global trade negotiations. *Development, 43*(2), 77–82.

Schafft, K. A., Jensen, E. B., & Hinrichs, C. C. (2009). Food deserts and overweight schoolchildren: Evidence from Pennsylvania. *Rural Sociology, 74*(2), 153–177.

Schlosser, E. (2001). *Fast food nation: The dark side of the all-American meal.* New York, NY: Houghton Mifflin.

Sherman, J. (2009). *Those who work, those who don't: Poverty, morality, and family in rural America.* Minneapolis: University of Minnesota Press.

Siefert, K., Heflin, C. M., Corcoran, M. E., & Williams, D. R. (2004). Food insufficiency and physical and mental health in a longitudinal survey of welfare recipients. *Journal of Health and Social Behavior, 45*(2), 171–186.

Slack, T. (2010). Working poverty across the metro-nonmetro divide: A quarter century in perspective, 1979–2003. *Rural Sociology, 75*(3), 363–387.

Swenson, D. (2009). *Investigating the potential economic impacts of local foods for Southeast Iowa.* Ames, IA: Leopold Center for Sustainable Agriculture.

Retrieved from http://www.leopold.iastate.edu/sites/default/files/pubs-and-papers/2010-01-investigating-potential-economic-impacts-local-foods-southeast-iowa.pdf

Tickamyer, A. R. (2000). Space matters! Spatial inequality in future sociology. *Contemporary Sociology, 29*(6), 805–813.

Tickamyer, A. R., & Duncan, C. M. (1990). Poverty and opportunity structure in rural America. *Annual Review Sociology, 16*, 67–86.

United States Department of Agriculture (USDA). (2007). *Total farm and farm sales.* US Department of Agriculture, Agricultural Census 2007, Congressional Districts, National Agricultural Statistics Services.

United States Department of Agriculture (USDA). (2009). *Access to affordable and nutritious food: Measuring and understanding food deserts and their consequences.* U.S. Department of Agriculture, Economic Research Service. Retrieved from http://www.ers.usda.gov/media/242675/ap036_1_.pdf

United States Department of Agriculture (USDA). (2010). *Farmers Market Growth: 1994–2009.* Washington, DC: Agricultural Marketing Service.

United States Department of Agriculture (USDA). (2012). *Farmers markets and local food marketing.* Washington, DC: Agricultural Marketing Service.

Vozoris, N. T., & Tarasuk, V. S. (2003). Household food insufficiency is associated with poorer health. *Journal of Nutrition, 133*(1), 120–126.

Welsh, R. (2009). Farm and market structure, industrial regulation and rural community welfare: Conceptual and methodological issues. *Agriculture and Human Values, 26*(1–2), 21–28.

Wilson, J. W. (1996). *When work disappears: The world of the new urban poor.* New York, NY: Knopf.

Contributors

Kaitland M. Byrd is a PhD student in the department of sociology at Virginia Tech. Her research areas are culture and the sociology of knowledge and power, with special emphasis on Southern foodways.

Kevin M. Fitzpatrick, PhD, is a University professor of sociology and Jones Chair in Community at the University of Arkansas. His research interests focus primarily on place-based analyses of health and well-being. In addition, he continues his work on disadvantaged populations with a particular focus on health and nutrition among low-income students and mental health among the homeless.

Meredith Gartin is a medical anthropologist who studies the human dimensions of global environmental health and urbanization, with a particular focus on vulnerabilities related to food and water. She holds a BA in environmental anthropology from the University of Georgia, an MA in sociology from Auburn University, a PhD in global health from Arizona State University, and has postdoctoral training from the Urban Sustainability Research Coordination Network (RCN), a NSF-funded research initiative established to examine sustainability as a unifying principle that connects cities worldwide in their transition toward equity, security, and justice.

Leslie H. Hossfeld is professor of sociology and chair of the Department of Sociology and Criminology at the University of North Carolina Wilmington. Dr. Hossfeld is trained in rural sociology from North Carolina State University College of Agriculture and Life Sciences. She has extensive experience examining rural poverty and economic restructuring and has made two presentations to the United States Congress and one to the North Carolina Legislature on job loss and rural economic decline. Dr. Hossfeld has served as cochair of the American Sociological Association Task Force on Public Sociology, vice president of Sociologists for Women in Society, and president of the Southern Sociological Society. She works on economic recovery projects for rural North Carolina counties and is cofounder and executive director of the Southeastern North Carolina Food Systems Program Feast Down East.

E. Brooke Kelly is an associate professor of sociology at the University of North Carolina Pembroke, where she worked with students and community partners as part of a social research methods course to collect some of the data on limited-resource farmers, addressed in the chapter. Dr. Kelly's teaching and research interests broadly focus on inequalities, work, and family. She has worked on numerous community based research projects that address poverty and food insecurity. Dr. Kelly has served as chair of the Poverty, Class, and Inequalities Division of the Society for the Study of Social Problems and is currently chairing the Southern Sociological Society's Committee on Sociological Practice. She has also served as a fellow and research affiliate of the Rural Policy Research Institute's Rural Poverty Center.

Janna Lafferty is currently a PhD candidate in the Department of Global and Sociocultural Studies at Florida International University, concentrating on critical food studies as an environmental anthropologist. Her dissertation research focuses on identities, inequalities, and the cultural politics of alternative food activism in the US Pacific Northwest.

Savannah Larimore is a graduate student in the Department of Sociology at the University of Washington. Her research interests include the cultural, social, and environmental factors contributing to unequal patterns in food access across social groups, as well as the role these factors play in health disparities. She is also interested in the factors contributing to the underrepresentation of women and minorities in STEM fields.

Kathleen LeBesco, PhD, is Associate Dean for Academic Affairs at Marymount Manhattan College in New York City. She is author of *Revolting bodies: The struggle to redefine fat identity*, co-author of *Culinary capital*, and coeditor of *Bodies out of bounds: Fatness and transgression, edible ideologies: Representing food and meaning*, *The Drag King anthology*, and several journal special issues. Her work concerns food and popular culture, fat activism, disability and representation, working-class identity, and queer politics.

Carlos J. Maya-Ambía studied Economics at the National Autonomous University of Mexico and got both a master's degree in Political Science and a PhD in Economics and Social Science from the Free University of Berlin.. He has been researcher-professor at several institutions in Mexico, the United States, Europe, and Japan. Currently, he is researcher-professor at the Department of Pacific Studies (University of Guadalajara). He has published five books as author, eleven books as editor, eighty-five articles, and nineteen book reviews in Mexican and international scientific journals. His current research interests are: the Japanese horticultural market; political economy of agriculture and food; and the critique of Karl Polanyi to the Market Society.

Marcel Mazoyer is honorary professor of comparative agriculture and agricultural development at AgroParisTech and visiting professor at the University of Paris XI. He has piloted numerous studies and research programs on the economics and development of farm-holdings, markets in farm produce, agrarian systems, and agricultural and rural development projects in addition to programs and policies in several African, Latin American, Asian, and European countries. He was previously research director and head of the Rural Economics and Sociology Department at the *Institut National de la Recherche Agronomique*, titular professor of the chair of Comparative Agriculture and head of the Department of Economic and Social Sciences at the *Institut National Agronomique Paris-Grignon*, guest professor at universities in several countries, and chair of the Program Committee of the FAO. Professor Mazoyer is the author of several books and numerous articles.

Peter Naccarato, PhD, is professor of English and world literatures at Marymount Manhattan College in New York City. He is coauthor of *Culinary capital* and coeditor of *Edible ideologies: Representing food and meaning*. His scholarly work is in the area of food studies, focusing on the role of food and food practices in circulating ideologies and sustaining individual and group identities.

Laurence Roudart is professor of agricultural development in the Faculty of Social and Political Sciences at the *Université Libre de Bruxelles*, where she occupies the chair devoted to The agrarian issue in developing countries. She heads the Master in Development Studies and is deputy director of the Institute of Sociology in this University. Her research and teaching focus on matters relating to agricultural policies, land policies, and food-security policies in developing countries. She has carried out fieldwork in Indonesia, Egypt, Senegal, Mali, and Burundi. She is a member of: the steering group of the editorial committee of the journal *Mondes en développement*, a member of the board of editors of the journal *World Food Policy*, and a consultant to *Encyclopaedia universalis* for articles dealing with agriculture or food security. She is also a member of the Scientific Council of the Michel Serres Institute for Resources and Public Goods.

Vaughn Schmutz is an assistant professor of sociology at University of North Carolina at Charlotte. He studies the dynamics of classification, evaluation, and consecration in a variety of cultural fields. His published work has appeared in *Social Forces, Poetics, Cultural Sociology, American Behavioral Scientist, Popular Music & Society, Tijdschrift voor Sociologie*, and *Sociologie de l'Art.*

Jennifer Sumner teaches in the Adult Education and Community Development Program at OISE/University of Toronto. Her research interests include food, sustainable food systems, cooperatives, and rural communities. She is the author of *Sustainability and the civil commons: Rural communities in the age of clobalization* and coeditor of *Critical perspectives in food studies.*

Amanda Smith holds an MA from the University of North Carolina Wilmington, where she currently teaches sociology. Smith's work focuses on fair and alternative trade, globalization, and food systems. Her past research endeavors have focused on examining the experiences of small farmers participating in local food systems and fair trade. Smith has worked extensively with small farmers and local food-systems organizations in southeastern North Carolina. Additionally, in 2012, she received the Ralph W. Bauer Fellowship, which supported her research examining fair-trade coffee farmers in the Matagalpa region of Nicaragua.

Mark B. Tauger is an associate professor of history at West Virginia University. After earlier work in musicology, he earned his PhD from UCLA and conducted research in Russia and Ukraine with support from IREX. He has published extensively on Soviet famines and agriculture, the Bengal famine in India in World War II, world agricultural history, and other subjects. His work has won the Eric Wolf Prize from the *Journal of Peasant Studies* and the Rasmussen Prize of the Agricultural History Society.

Andria D. Timmer is a lecturer of anthropology at Christopher Newport University. Her research examines patterns of inequality and the efforts of individuals and groups to make social change. This focus has led her to explore childhood malnutrition in Central America, educational inequalities in Hungary, and the food movement in the United States.

Julia F. Waity is an assistant professor of sociology at the University of North Carolina Wilmington. Dr. Waity's research focuses on food insecurity and spatial inequality. She has presented her work at the American Sociological Association, the Society for the Study of Social Problems, and the Research Innovation and Development Grants in Economics (RIDGE) Conference. She has received funding for her research from Indiana University, where she earned her PhD, the University of North Carolina Wilmington, and the Southern Rural Development Center.

Susanne A. Wengle is a lecturer in the Political Science Department at the University of Chicago. She holds a PhD in political science from the University of California, Berkeley. Her research investigates the governance of markets. Her empirical and theoretical focus is on how polities forge and adapt market institutions and how market outcomes in turn shape the political arena. She has authored a book on Russia's post-Soviet market transformation, *Post-Soviet power*. Her articles have appeared in *Governance & Regulation*, *Studies in Comparative International Development*, *Economy and Society*, *Europe-Asia Studies*, the *Chicago Policy Review*, and the *Russian Analytical Digest*. She is currently working on a project on the political economy of agriculture and food in the United States and Russia.

Don Willis is a PhD student and Huggins Fellow at the University of Missouri, Department of Sociology. He received his MA in sociology from the University of Arkansas. His research interests include health disparities and their overlap with food and social capital. In particular, his work highlights the influence of social context on food insecurity and its associated health outcomes.

Index

Printed and bound by CPI Group (UK) Ltd, Croydon, CR0 4YY